Event-Database Architecture for Computer Games

Event-Database Architecture for Computer Games proposes the first explicit software architecture for game development, answering the problem of building modern Computer Games with little or no game design. An archetypal software production process, based on this architecture, is also introduced.

This volume begins by describing the formal definition of software production processes in general and the production process of Computer Games in particular. It introduces the two basic principles behind the software architecture that addresses the communication and productivity problems of a degenerative production process. It goes on to describe the archetypal software production process and outlines the role that the Game Designers, Game Programmers, Game Artists, Sound Designers and Game Testers play in that process.

This book will be of great interest to professional game developers involved in management roles such as Technical Directors and Game Producers and technical roles, such as Tools Programmers, UI Programmers, Gameplay Programmers and Engineers, as well as students studying game development and programming.

Rodney Quaye is Senior Software Development Engineer in Test at Build A Rocket Boy. He has worked in the Computer Games industry for over 16 years. He has worked at several Games Studios including Sumo Digital, nDreams, Supermassive Games, Traveller's Tales, Hotgen, Oysterworld, Second Impact, Flaming Pumpkin, Goldhawk Interactive, Jagex, Gusto Games, Criterion, Asylum Entertainment, Codemasters and Deibus Studios. The famous titles he has worked on include *Burnout 2 and 3* for Criterion, *LMA Manager* for Codemasters, *Runescape* for Jagex, *Lego Worlds* for Traveller's Tales, and *Everywhere* for Build A Rocket Boy.

Event-Database Architecture for Computer Games

Volume 1, Software Architecture and the Software Production Process

Rodney Quaye

CRC Press
Taylor & Francis Group
Boca Raton London New York

CRC Press is an imprint of the
Taylor & Francis Group, an **informa** business

Designed cover image: Shutterstock

First edition published 2026
by CRC Press
2385 NW Executive Center Drive, Suite 320, Boca Raton FL 33431

and by CRC Press
4 Park Square, Milton Park, Abingdon, Oxon, OX14 4RN

CRC Press is an imprint of Taylor & Francis Group, LLC

© 2026 Rodney Quaye

ISBN: 9781032820675 (hbk)
ISBN: 9781032818061 (pbk)
ISBN: 9781003502784 (ebk)

DOI: 10.1201/9781003502784

Typeset in Times
by KnowledgeWorks Global Ltd.

Access the Support Materials: www.routledge.com/9781032818061

Contents

About the Author

Rodney Quaye is Senior Programmer who has worked in the Computer Games industry for over 16 years. He was born in the UK but grew up in his fatherland, Ghana, attending primary school there. He returned to the UK to attend secondary school. He grew up playing Computer Games at school and university but never thought of it as a career. He graduated from the University of Warwick with a Bachelor of Engineering degree in Computer Systems Engineering in 1993. He went to work as a programmer first on medical information systems for hospitals and then market analysis systems, mainly for car manufacturers. He then had a near-death experience which gave him a spiritual awakening. He reflected on his life and realised that his heart was not in his work. He felt God was calling him back to his first love, Computer Games. So he started a career in that industry in 1999, working at several Games Studios including Sumo Digital, nDreams, Supermassive Games, Traveller's Tales, Hotgen, Oysterworld, Second Impact, Flaming Pumpkin, Goldhawk Interactive, Jagex, Gusto Games, Criterion, Asylum Entertainment, Codemasters and Deibus Studios. The famous titles he has worked on include *Burnout 2 and 3* for Criterion, *LMA Manager* for Codemasters, *Runescape* for Jagex and *Lego Worlds* for Traveller's Tales. He wrote this book to provide a standard documented software architecture for making Computer Games.

Introduction

In this volume in this series, ***Event-Database Architecture for Computer Games: Volume 1, Software Architecture and the Software Production Process,*** the problem of building modern Computer Games with little or no game design will be introduced, along with a software architecture for solving this problem.

An archetypal software production process, based on this architecture, will also be explained.

1 The Problem

The *classic software production life cycle*[1] is meant to begin with the analysis of the problem the software would address, followed by the drafting of its manual, its design and a plan to test it. Following these initial theoretical phases are two more practical ones. These include the implementation of the design of the software, and the testing of that software, based on the plan drawn up earlier, respectively.

The analysis of the problem would include a detailed description of the requirements which the software will meet. These would be either drawn up in consultation with a customer, for whom the software would be written for. Or these would be drawn up after market research had been carried out on a consumer market. The resultant document, drawn up from this analysis, is known as the User Specification. This document acts as a basis for the next phase; that is the drafting of the manual that would accompany the software.

On some occasions, however, before this phase, the analysis may also act as a basis for the study of the feasibility of the software. This would assess the computer hardware it was for, the time and the tools available to make it. And the results of this study would be included in the User Specification.

On other occasions, this feasibility study may be done after the software has been designed, to give a more accurate study. The *software design*[2] includes a breakdown of the *software modules*,[3] the *software data*,[4] the *software library*[5] and other tools that would be used to meet the requirements. If the feasibility study were conducted after the *software design*, then it would also be included in that design. The *software design* acts as a basis for the next phase that is the drafting of the initial plan to test the software.

After the plan for testing the software had been drafted, the tools chosen in the *software design* would be used to build the components of the software and assemble these together. Finally, the result would be tested against the *software design*, according to the plan.

There are many variations on this process. Some repeat the cycle, beginning with the analysis and ending with the testing, several times. Others repeat only some of the phases of the cycle several times. Some repeat the cycle to produce each of the software components. But if you at least understand the *classic software production life cycle*, you can understand the different phases of these other processes.

The production of a Computer Game too follows this basic pattern. That is, it begins with an analysis of the requirements of the software. It begins with a general *game design*.[6] This is equivalent to a User Specification. The document includes a description of the goals of the game, the different stages, the goal of each stage and the progression through the stages. It includes a description of the different items in each stage, how these appear and behave. It also includes a description of the *User Interface*[7] and how this would be used to interact with the items in each stage, to progress through the game. Furthermore, the analysis typically includes a study of

DOI: 10.1201/9781003502784-1

the feasibility of the software. This involves a small sample of the software being written and built, to give a practical demonstration of a short, but important, part of the game.

After the *game design*, a second design is written known as a *technical design*.[8] This includes a description of what *data* would be needed to build the game, and the tools required to create or manage that *data*. It also includes a breakdown of the *software modules* (or *game modules*[9]), and the *software library* (or *game-engine*[10]) needed to implement the *game design*, using that *data*.

Following the *technical design*, the various staff responsible for building the game set about getting or creating the *data*, the tools, the *game modules* and the *game-engine*. They assemble these together to build the game. After that, they draw up a plan to test the result against the original design. They execute that plan. And, finally, they draw up a *User Manual*[11] for the game.

The only major difference between this production process and the *classic software production life cycle* is cosmetic. Namely, all the phases of the *software production life cycle*, which occur throughout the process, fall into three broader phases known as *Pre-production, Production and QA*.[12] Each of these phases incorporates two or more phases of the *software production life cycle*. *Pre-production* incorporates four phases: the analysis of the requirements, the documentation of these in the User Specification, the drafting of the *software design* and the study of its feasibility. *Production* incorporates four phases: the re-analysis of the requirements, the re-drafting of the User Specification, the re-drafting of the *software design* and the implementation of that design. *QA* incorporates three phases: the drafting of the manual for the software, the planning of the final test and the execution of that plan. Apart from this difference, the process begins with a plan, and it follows with the implementation of that plan; just like the *classic software production life cycle*. There is a diagram showing the steps of the *classic software production life cycle* in Figure 1.1. There is a diagram showing the steps of the production process of Computer Games in Figure 1.2.

At least, this is what is meant to happen. But, instead, what actually does happen is something else. The desire to get funding for the Computer Game distorts the entire production process. The analysis of the game, which normally proceeds the process, is hastily put together. The *game design* (i.e. the User Specification) is anything but specific. It is vague and incomplete. It contains just enough highlights to sell the game to whoever is funding the project.

The financial backers of the project may just want to see a summary of the details of the game. Or they may want to reserve the right to change aspects of it, at a later date. These options do not stop the *Software Developer*[13] from completing the *game design*, with details which may later be changed. Nor do these options prevent the *Developer* from building software, to support this completed *game design*, and trying to re-use the tools later on, when the *game design* had changed. Nor do these options stop the *Developer* from presenting a summary of this completed *game design* to the financial backers. However, it is not in the *Software Developer*'s interest to commit to work which they may have to change later on. It is less expense for them to leave that work out completely.

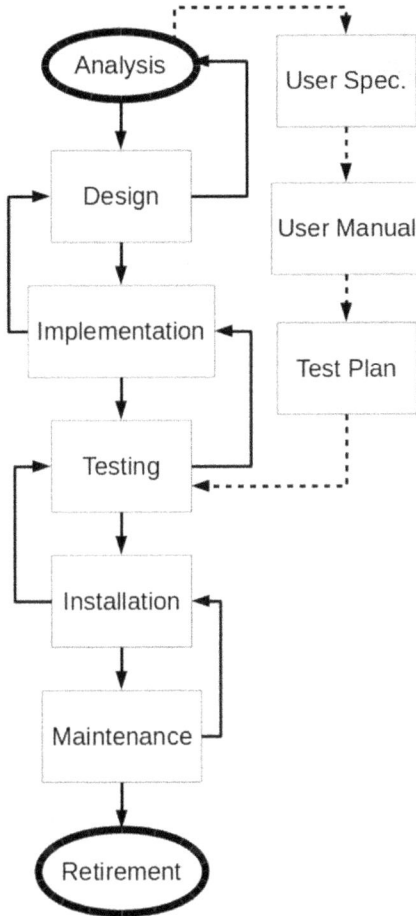

FIGURE 1.1   The classic software production process.

So just like the User Specification, the *Software Developer* similarly neglects the feasibility study. They rush it through with optimistic speculations. It invariably produces the positive result that the *Software Developer* requires to secure the funding. The software that is written to demonstrate the highlights of the game has just one aim. And this is the sale of the game to the financial backer. The *Software Developer* has no intention to re-use it. And the direct result, of the rush through this initial phase, is that the rest of the *game design* has to be made up after the *technical design* has been written. That is, the completion of the *game design* has to occur during the building and the assembly of the software components.

But before that phase, the *technical design* is written on the basis of the incomplete *game design*. And it too is driven by the desire to secure funding. Thus, the choice of any *game-engine*, in the *technical design*, is tactless. The *game-engine* is chosen for its ability to help build games on many different *platforms*[14]; especially

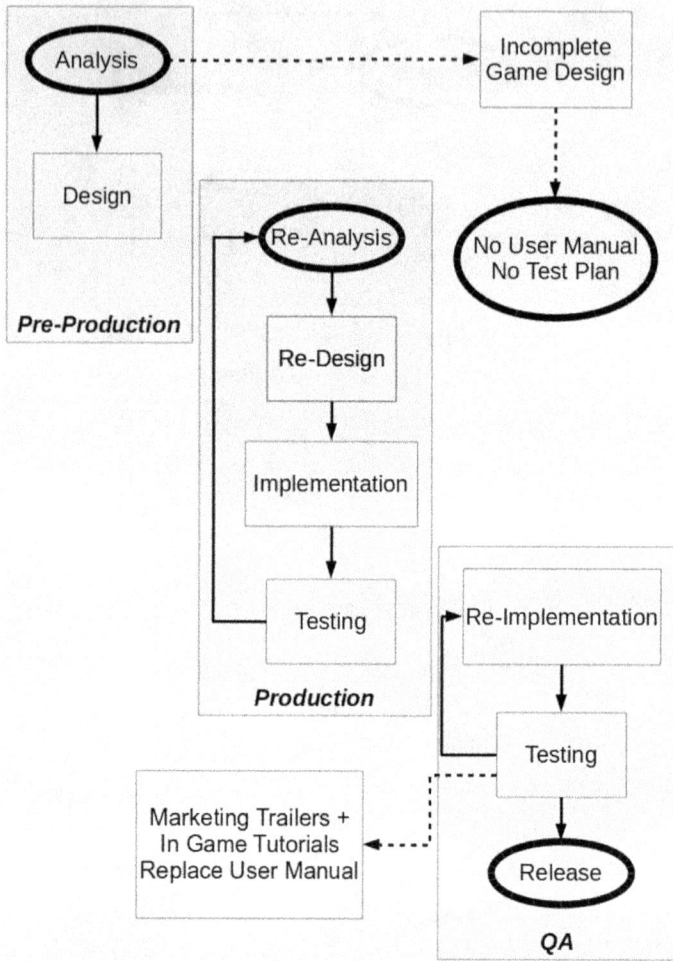

FIGURE 1.2   The production process of the Computer Games industry.

the latest *platform*. This in turn helps promote the project to any financial backers interested in new lucrative markets.

Likewise, any tool that could be used to edit elements of the game, such as a *game-editor*,[15] is chosen for its ability to seemingly provide the project with options. The more that it appears the project could handle changes in the *game design*, the less risk it appears to have for any financial investor. Neither the *game-engine*, the *game-editor*, nor any other tool would be chosen for its ability to meet the requirements of the *game design*. Since these requirements were not completed anyway, all of the choices in the *technical design* are makeshift ones. And these are meant to be partially, if not completely, discarded later on, during the rest of the production process.

After the *technical design* has been written, the production of the software components begin. But the effects of the absence of so many provisions, in the *game design* and *technical design*, on the *software data*, tools and *modules* are disastrous. When it comes to building and assembling the software components, all of these are continuously being rewritten, as informal changes are being made to the *game design*. Each change, and subsequent pass through the *software production life cycle*, is dealt with less rigorously than the preceding one. And the production process rapidly descends into a reactionary, instinctive and primeval state.

Eventually, the time between each informal proposal for some change, and its appearance in the game, is a matter of hours, minutes and seconds. The *Software Developer* completely neglects the *game design* and *technical design*. The ad hoc process that emerges from this descent is known as the *Software Evolution Process*.[16]

1.1 THE SOFTWARE EVOLUTION PROCESS

Although they stumble into it every time, most *Software Developers* do not recognise the *Software Evolution Process*. They merely react to it when it occurs. But others do recognise it. And they actually advocate it to their financial backers as a sound, if not scientific, method for developing the incomplete *game designs* which they produce. To them, it represents a *practical application*[17] of the theory of Biological Evolution, to the problem of quickly responding to the wishes of their financial backers, and their changes to the *game design*.

This is despite the fact that, in many ways, the *Software Evolution Process* does not bear any resemblance to the theory of Biological Evolution. The theory of Biological Evolution requires small mutations in animals, which allow the animals to slowly adapt to the environment. The *Software Evolution Process*, on the other hand, involves coping with massive, traumatic changes in a *game design*, and getting an instant response from the production process. The theory of Biological Evolution requires a period of time during which two or more mutations live alongside each other, until the environment eliminates one. Indeed, some mutations may continue and proliferate into different species. In contrast, the *Software Evolution Process* ensures that one version of a game quickly supersedes another. And that at no time are there two or more versions in production. The theory of Biological Evolution involves a slow, steady process of change which always leads to progress. The *Software Evolution Process*, however, involves a fast, volatile process of change, which leads just as much to progress, as it does to regress.

Very few aspects of the *Software Evolution Process* resemble the theory of Biological Evolution. And those that do are not commendable. The theory of Biological Evolution involves a long, *open-ended process*,[18] which develops over millions and billions of years. Similarly, the *Software Evolution Process* feels just as long, it is just as ageing and just as pointless. The only difference being that the *Software Evolution Process* manages to cram the entire exhausting experience into a few months or years. The theory of Biological Evolution is hard to prove, because of the vast time frame that it requires. You can observe the individual mechanics of it (the mutations, how changes of environment affect the survival of animals). But you will not live long enough to observe these working progressively in nature. Similarly, the *Software Evolution Process* is hard to prove or disprove. Although you can quickly observe some of the effects on a game, that arises from each informal change that is made, you have no idea what the long-term effects will be.

The reason for this is that the *Software Evolution Process* relies on overlapping phases of the *software production life cycle*, and ignoring documentation, to quickly react to changes. It does not only overlap the *game design*, with the building, of the software. It also overlaps the feasibility, the *technical design* and testing of the software too. While one group is proposing informal changes to the *game design*, another is assessing the feasibility of a different set of changes, another is changing the *technical design*, another is building components of the software and another is testing a previous set of changes. This makes the *Software Evolution Process* both very complex, and at the same time unscrutable.

Yet this practice has become so commonplace in the Computer Games industry, it has become formalised. Such that, for example, prospective *Game Artists*, who attended formal interviews at *Software Developers*, have been expected to have some prior experience designing and running a distinct sub-process, during production process, known as an *Art Pipeline*.[19] The phases of which would run parallel to the phases of the *software production life cycle*, that the rest of the staff would follow. And product of which would be all the special effects, the characters, the animations, the locations, the menus and other graphical items in the game. In other words, through the *Art Pipeline*, the *Artists* have been expected to design and build all of these components of the *User Interface*, at the same time the rest of the game was being developed by other staff.

Likewise prospective *Game Programmers* known as Build Engineers who attend formal interviews at *Software Developers* have been expected to have had prior experience running a distinct sub-process known as the *Build Pipeline*.[20] Again the phases of this would run parallel to the phases of the *software production life cycle* that the rest of the staff would follow. And the product of which is the latest version of the game built by an automated system, on a daily or hourly basis. To ensure that there were no errors building and testing the game after changes had been introduced to the *game design*. These changes occur at a rapid rate, 80–100 changes every day, 3–4 changes every hour and 1 change in every 15 minutes.

Nevertheless, the effect of having so many sub-processes, such as the *Art Pipeline* and *Build Pipeline*, running concurrently in the *Software Evolution Process*, offers the *Software Developer* an advantage. Although this makes the process very complex, this also makes it unscrutable. And this is why some *Software Developers* advocate it to their financial backers: not for its responsiveness, but its inscrutability.

On the one hand, the *Software Developer* advocates the expediency of the *Software Evolution Process*, for rapidly coping with changes to a *game design*, when production begins. On the other hand, after errors have occurred, the *Developer* falls back on the complexity (and inscrutability) of the process to account for these errors. On the one hand, the *Developer* claims to appreciate the need to address any demands, from the financial backers, for changes to the *game design*. On the other hand, the *Developer* has no qualms about adopting a process which neither they, nor the investors, can understand. How can they appreciate the suitability of a process which seems to begin well enough, but they always lose track of? Their lack of understanding is self-evident when you ask them to define what the *Software Evolution Process* is.

No two *Software Developers* have exactly the same idea of what the *Software Evolution Process* is. The degree to which they overlap the phases of the *software production life cycle*, which phases they repeat, which phases they leave out, and the number of sub-process running in parallel within the *Software Evolution Process* differs. The same *Developer* never uses the same *Software Evolution Process* twice. This is consistent with the inscrutability of the *Process*. When it ends, the *Process* is inscrutable and the *Developer* cannot analyse the phases of the *Process*. They cannot identify which phases were successful and which phases failed. Therefore they cannot identify which phases to repeat and which phases not to repeat. The *Process* is inherently ad hoc. It changes as the need arises and is therefore unrepeatable. The *Software Evolution Process* is phenomenon which the *Developers* find expedient but mysterious.

But if they looked backed into its history, they would find out that there was no great mystery about its origins. The *Software Evolution Process* is older than the Software industry, and much older than the Computer Games industry. It was the process which would have been used to produce the first-ever piece of software. Back then as now, the two characteristics of a *Software Evolution Process* were meant to be in theory, firstly, that it slowly evolves and grows software over time. And secondly that the basis of this evolution was feedback from the software user. But, in practice, the process has never managed to *evolve software without degenerating*.[21] Nor has *feedback from the software users*[22] played any part in it. Instead the feedback has come from staff involved in the software production process who act as proxies for software users. In the Computer Games industry, the *Game Producers*,[23] *Game Designers*,[24] *Game Testers*[25] and financial backers have traditionally taken up this role upon themselves. The financial backers especially provide the initial feedback which secures the investment when the production of the game begins. And this feedback, from the financial backers, continues to secure the financing from the beginning to the just before the end of the process.

So the characteristics of a *Software Evolution Process*, such as they are, have always been external characteristics related to why the process was being used, not internal characteristics related to how it was being conducted. That is, firstly, the process was being used because of the absence of an initial plan to build the software, in this case a Computer Game, before the date it was scheduled to be released. And the need to fill the void due to this absence. Secondly, the process was being used because of the expediency of overlapping some phases of the *software production life cycle* (e.g. feasibility study, *game design*, *technical design*, implementation and so on) and omitting others (e.g. a User Manual and a test plan). To speed up the production process and produce visible results as soon as possible. And reassure the financial backers and the rest of the staff.

For the *Software Evolution Process* is only a default process. It is the process which someone, who has never made a Computer Game before, would use to produce one. It is the process which any *Software Developers* would have to adopt, regardless of their experience or knowledge, in the absence of any planned approach to build a Computer Game. It fills the gap between the lack of a decision about a *game design* and the final product. This is a gap which those *Software Developers* who advocate the process hope will be very long. So long that, at the end of the process, no one will remember or even care about the beginning. Any process which falls into this gap, and is mutable, qualifies as a *Software Evolution Process*.

None of the *Software Developers* who espouse the virtues of the *Software Evolution Process*, to their financial backers, would go into its effects in detail. For this would be invariably damaging. The process repeatedly destroys any definition which it acquires. With each change introduced into the *game design*, the process itself mutates. And the phases of the process which overlap, how much these overlap, the phases left out and the number of sub-processes running within the process changes. These mutations destroy the definition within many different parts of the project. Not least amongst these are the definitions within the software itself.

1.2   THE EFFECT ON SOFTWARE

The damage the mutating process does to the definition within the software is critical. The *game modules* and *game-engine* pull all the resources together to build the game. The importance of these two pieces of software inflates, during the *Software Evolution Process*. For these are responsible for the only remotely comprehensive documentation, which is kept up to date. Namely, the computer files, written by the *Game Programmers*,[26] which they use to build the *game modules* and *game-engine*. These files are only comprehensible to the *Programmers*. And the *Game Producers* and *Designers* rely on the translations of these files, by the *Programmers*, to understand the current state of the software. They also rely on feedback, from this software, to make up the rest of the *game design*. But although the *game modules* and *game-engine*, like all computer programs, are written to cope with well-defined behaviours, the *Software Evolution Process* destroys any definitions these acquire as it mutates. Instead, during the process, the *Producers* and *Designers* like to think in terms of events, which are infinitely interchangeable.

When an event *A* happens, event *B* should follow it. Or event *B* should follow a set time after *A*. Or *C* should happen the same time as *A*. For example, in the context of a game based on managing football clubs, event *A* may be the manager of a club fining a player for misconduct. Event *B* may be an article appearing in a newspaper, speculating about the transfer of a player from one club to another. And event *C* may be a request from that player, to the manager of the club, to be transferred.

In the context of a game based on racing cars, event *A* may be the start of a race. Event *B* may be a car revving up its engine. And event *C* may be the sound of a crowd cheering.

Some of these events may already be defined in one context and produced by the *game-engine* or the *game modules*. Others may not even exist and may just be a coincidence of different features added to the game.

For example, in the game based on football just mentioned, the article that would appear in the newspapers may only be originally defined in the context of the day after a match, involving a football club. And the text of the article presumed a match had just been played and was merely meant to be complimenting a football player's good performance in that match. It was never meant to be a response to that player being fined by the club's manager. Nor was it meant to be a prelude to the player requesting a transfer because of such a fine.

Until that is, one day, through a coincidence, a *Game Producer* or *Game Designer* happened to notice an article appear in the newspapers, complimenting a football player's performance after a match, while casually playing the Computer Game. And coincidentally they subsequently noticed that another player from the same match had been sent off for violent conduct. From these two coincidences the *Producer* or *Designer* has an epiphany. They presume that the articles in the newspapers were responding to what was happening in the football match. And that this mechanic could be extended to make the articles also respond to what was happening at the football club. Whenever a player was disgruntled with the club's manager for one reason or another, at any point in time in a football season.

So the *Producer* or *Designer* immediately decide, that same day, to change the *game design* and add this feature. Even though this mechanic or relationship between the events on the football match and the newspaper articles was incidental and did not really exist. Even though, in the original context, the newspaper articles were only meant to compliment one football player's good performance in a match and only appear one day after that match.

Another example, in the game based on racing cars, would be the sound of a car revving up its engine. Originally, this sound may just be a coincidence to do with the mechanics of the car. It may just be a coincidence of the motors accelerating, in a high gear, and the clutch being disengaged after some accident, which a *Producer* or *Designer* happened to notice. If the car had not been damaged, then it would not be making that sound. Neither was it the intention for this to be heard at the start of a race. Since the clutch would presumably not be damaged. And it would be impractical to get a quick start with it manually disengaged, and the motors revving in a high gear.

However, when the *Producer* or *Designer* noticed this sound, they had an epiphany. What if this sound could be heard at the start of the race? What if all of the engines at the start were making this sound on the start grid? It would really build up the tension and get the players excited.

So again the *Producer* or *Designer* immediately decide, that same day, to change the *game design* and add this feature. Even though this relationship between the sound of the motors revving after an accident and the sound the cars made at the start of a race was incidental and did not really exist. Even though, in the original context, this sound was only generated by the *game-engine* as it tried to simulate the physics of a car after a crash.

When the *Producers* or *Designers* see an event in the game, they presume that it can organically evolve. That event can be taken from one context and put in another. Or they presume it can be extended or modified slightly. When, in fact, that event does not really exist and is a coincidence of two or more other events, or that event has already been tightly coupled within a well-defined behaviour. Thus, the accumulative effect of each daily change to the *game design*, and subsequent modification of the software, during the *Software Evolution Process*, after the *Game Producers*, *Game Designers* and *Game Testers* play the game and give feedback, beginning with the initial *game design*, is a slow and steady a loss of definition in the software.

Also during the *Software Evolution Process*, the features which they add to the *game design*, after they give feedback, often overlap. But these overlaps do not become apparent until much later on, during production process. Since there is little or no analysis of the product at the beginning. These overlaps only emerge as technical problems, which occur when writing the *game modules*, or building the game, or when testing the software at the end of production.

For example, when the manager of a football club wins their first match in a season, a *Game Producer*, working with one group of staff, may propose that a unique article should appear in the media. The sports journalist should make sarcastic comments about the surprise victory for the newly appointed, previously unheard of, manager; that is the player of the Computer Game. And thus tease that player who takes over the role of a football manager, when the game begins, to make him or her relax. But a *Game Designer*, working simultaneously, with a completely different group of staff, may propose shortly afterward, that the football manager should also receive an article in the media whenever they win a trophy. And this article should congratulate and praise the manager for the club's achievement.

Now suppose the game follows the football seasons in many countries. And in some countries the first match is the final of a competition. This match acts as a grand opening, or curtain raiser, to the coming football season. In such a season, both proposals, from the *Game Producer* and *Designer*, overlap and produce a contradiction. On the one hand, the manager who won the first match would receive condescending remarks, from journalists, for a lack of experience. On the other hand, on the same day, the manager would also receive praise, from journalists, for winning a competition. This contradiction would only become apparent when either the *game modules* were being written and built, or the game was being tested much later on.

When the contradiction emerges, typically in the *Software Evolution Process*, this does not cause a re-analysis of the product. More so if it emerges a long time after the production begins, especially towards the end; during the nominal *QA* phase of the production process or the testing of the game. At this late stage, the authors of both proposals to change features of the *game design* tend to defend their choice and are reluctant to drop their proposals for someone else's. The *Game Producers*, *Designers* and other staff who added these informal changes to the *game design* view such actions as a waste of their time and effort: which it is. And, of course, there is no written comprehensive *game design* to help resolve the issue. So the compromise is usually to somehow keep both proposals. This is done by either giving a higher priority to one proposal over the other, during some periods of the game, and swapping the priorities during other periods. Or this is done by randomly choosing between the two proposals when they overlap during the game. Or this is done by giving a higher weighting of one proposal over another, during this random choice. So that it always occurs more frequently in the game than the other.

Thus the effect of the overlapping features of a game, which emerge late during the *Software Evolution Process*, is a loss of certainty. A feature which appears frequently at the beginning of production can suddenly seemingly disappear. And the staff cannot be certain whether that feature is missing either because of an error. Or whether it is missing because it had been given a lower priority than another feature. Or whether it is missing because it had been limited to certain periods of the game. Or whether it had been dropped all together.

1.3 THE EFFECT ON LANGUAGE

Nevertheless, even the loss of certainty amongst the staff, and the loss of definition, within the *game modules* and *engine*, is merely a symptom of the main effect of the mutations, of the *Software Evolution Process*. The main effect is the confusing language which those who follow it use. This stems from the fact that, as the *game design* changes, the *Game data* loses its definition. In the beginning, some of that *data* would have been well-documented in the initial *game design* and *technical design*. But as time passes, the use of some of the *data* becomes more extensive or, in other cases, very limited. The description of the *data* does not keep up with these changes. So the quality of the description begins to vary quite rapidly.

This quality of each description depends on when that piece of *data* was introduced. Those descriptions written at the beginning of the process would be more accurate than those written towards the end. The quality of the descriptions also depends on how quickly certain changes were required. And it depends on the group who made these changes and their knowledge of the history of the process.

The varying of the quality of the description, of the *Game data*, affects the *game modules* and the *game-engine* as well. If any *data* was well-defined and described, it would be easy to understand any *software module* which used that *data*. Even if the *module* were not well-documented, from the knowledge of the *data* that it used, and the purpose of the *module*, you could deduce how it functioned. But if the *data* was not well-defined, then this would not be possible.

Therefore the effect, of the poorly defined *Game data* on the *game modules* and *game-engine*, in a *Software Evolution Process*, is disastrous. The computer files, written by the *Game Programmers*, which uses the *Game data* to build this software, grow more and more mysterious. Eventually it becomes as much of a mystery to them, as the *Software Evolution Process* is to the *Software Developer*. Even the *Abstract data* (i.e. internal *data*) which they produce feels the effect of the lack of definition, of the *Game data*. The great majority of different types of *data*, which the *Programmers* use to write the software, are *Abstract data*. But these are merely derivatives of the general, *Game data*: these all share some relationship with *Game data*. And when the *Game data* has been poorly defined, the *Abstract data* also lack definition.

If the *Game data* were well-defined, then it would give the *Game Programmers*, who make up the *Abstract data,* a language to describe themselves. But when there is no language, the quality of description of the *Abstract data* varies from *Programmer* to *Programmer*. Depending on the extent of each one's knowledge, each *Programmer* makes up a name to describe each *data*. Rarely, some of them make an attempt to include additional explanations for the *data*. In general, however, most of them realise that this is futile, in light of the changing *game design*, and they do not bother. They rely solely on the single name they give each piece of *data* to describe it. But these names are arbitrary.

Since the *Game data* does not provide the *Programmers* with any language to describe the *Abstract data*, they reach for the nearest thing at hand. They use the very obscure, esoteric language that comes from either the various *game modules* or the *game-engine*, written by other *Programmers*. Or they use the language from their education, the latest technical articles they have read, or the tools they see around them. Or they use the language from the last game which they have played. They do not use a natural language that comes from the *game design*.

This same phenomenon repeats itself, simultaneously, in the other contingents, amongst the staff following the *Software Evolution Process*. But each occurrence of the phenomenon produces subtly different results. All the staff who produce the *Game data*, which the *Programmers* use, develop their own esoteric language. The *Game Designers*, the *Game Artists*[27] and the *Sound Designers*[28] all make up arbitrary names to describe the *data* they produce. And, later on, towards the end of the process, the *Game Testers* too wade in with their own language. But the esoteric language of each of these contingents differs slightly from the *Programmers'*, and from each other. The differences arise from the fact that each contingent does not share the same education, the same background, technical interests, or tools. Neither do they play the same games. Thus the combined effect, of each of these contingents, using all these languages, to communicate in a *Software Evolution Process*, is like the *Tower of Babel*.[29]

In biblical times, in the place which later became known as Babylon, man attempted to construct a tower to reach the heavens. But God stopped the construction of the *Tower of Babel* by causing all the builders to speak different languages. Likewise, in modern times, in the process which is known as the *Software Evolution Process*, a *Software Developer* tries to accomplish an ambitious project, without any vision, with the same results. Except this time, the process itself hinders construction by causing all the builders to speak different languages.

The overall effect is that the quality of communication within a *Software Evolution Process* degenerates over time. So much so that two *Game Artists*, two *Game Programmers*, two *Sound Designers* or two *Game Designers* frequently engage in their own *private conversations*.[30] Meanwhile, the rest of the staff stand idly by. These conversations may go on for several minutes in a meeting between the staff. Or sometimes, these may go on for days, or for weeks over several meetings. All the time, the rest of the staff present at these meetings look on, unable to comprehend, let alone contribute.

A lot of these *private conversations* occur through *memorandum*[31] in the form of E-mails. Take for example this E-mail:

```
'Fyi, we sometimes have poorly authored Logan strings with
missing token indices, and it's important to understand how
these should be handled when adding strings into the code.
There may have been emails sent around about this in the past,
but I can find them - so here's a new example.
   A made up news item with 3 alternative phrases:

   String_Id.sch#1  ""#4-Number# year old #2-Player# has
      suffered #3-Injury#.""
   String_Id.sch#2  ""#4-Number# year old #2-Player# was
      injured in the fixture against #7-Club#.""
   String_Id.sch#3  ""#2-Player# has suffered #3-Injury#.
      Manager #6-Staff# says he'll be back playing for
      #8-Club# in no time.""
```

The set of tokens used across all the alternatives is as follows:

```
2-Player
3-Injury
4-Number
6-Staff
7-Club (other club)
8-Club (player's club)
```

Not all token indices have been used (there's no token 1 or 5). Part of the text exporter process will recognise this and will renumber the tokens, e.g. as if the strings had been authored as follows.

```
String _ Id.sch#1  ""#3-Number# year old #1-Player# has
   suffered #2-Injury#.""
String _ Id.sch#3  ""#3-Number# year old #1-Player# was
   injured in the fixture against #5-Club#.""
String _ Id.sch#2  ""#1-Player# has suffered #2-Injury#.
   Manager #4-Staff# says he'll be back playing for
   #6-Club# in no time.""
```

However, the simplest way to think of this (rather than how tokens are renumbered) is that the token order specified in a call to NMAddNewsItem needs to match the numeric order in the string. So to add this news item you might implement it as follows (assuming a headline containing just the Player token):

```
NMSetParams8 (NM _ RESET, ""tytmStClClNuPlIn"", tNewsItem::
   NEWS _ TYPE _ XXXX, Club,
NIStaff(Manager),
NIClub(Club),
NIClub(OtherClub),
etc.
NMAddNewsItem (""Pl.PlInNuStClCl"", NEWS::THE _ HEADLINE _
   FOR _ STRING _ ID   <news::THE _ HEADLINE _ FOR _ STRING _
   ID>, NEWS::STRING _ ID <news::STRING _ ID>);
```

Bear in mind that strings entered into DevStrings.cpp won't have been renumbered (as it's one of the exporter tools that does this). This means that the above code would produce a broken news item until after a text export. This isn't really a problem so long as you're aware of it (and most of our strings are correctly authored and so don't have this issue).

If you have a poorly indexed string that you'd like to display correctly while it's still in DevStrings, I'd suggest manually renumbering it in DevStrings. If the string has alternative phrases but is already in Logan, you only really need to copy a single phrase into DevStrings and edit that.'

Source: A typical E-mail from a Software Evolution Process of Slippery Games Inc. Anonymous. June 2006

Now some of you may not understand this E-mail. But that does not matter. What does matter is that you do understand why this E-mail was written. And that when you understand why it was written you will understand what is wrong with it.

This E-mail is from a *Software Evolution Process* developing a game about managing football clubs or a football management game. Now if you have played these genre of games before, then you may recognise some of the words in this E-mail, and that the source of this E-mail was a production process for a football management game. But if playing a football management game were not part of your education or background or the types of games you have played in the past, then this would not be obvious. As has already been said, in a *Software Evolution Process* in the Computer Games industry, the staff will come from different backgrounds. And you will get some staff who have experience playing some genre of games (e.g. football management games), and others who have not. And for those who have not this E-mail would just be mysterious.

Now even those who have played this genre of games cannot tell the exact subject of this E-mail. It is full of vague acronyms e.g.

"NM",
"NI"
"Fyi"

It is full of cryptic abbreviations e.g.

"ty"
"tm"
"Pl"
"Cl"
"sch"

It is full of confusing keywords e.g.

"Logan"
"DevStrings"
"text exporter"
"exporter tools"

None of these words have anything to do with the football management genre. And even if you have played these games, then that experience is not going to be enough to help you understand what these words mean.

As has already been mentioned, it often seems that the *private conversations* that occur in a *Software Evolution Process* is occurring between highly skilled parties. But the author and recipient of this E-mail do not have any superior knowledge to the rest of the staff or anyone who has previously played a football management game before. The subject of the E-mail is a tool that allows you to format the headlines of news articles that appear during the football season when playing the game. The articles announce the injuries that have occurred to football players during a football match. That is it! As you can see, this is not a tool meant to be only used by highly skilled staff.

The tool was meant to be used by *Game Producers*, *Game Designers*, *Game Testers*, *Game Programmers* and even the software user or player. To add features to the game or to correct features in the *game design*. But the language describing that tool has degenerated to the point where only a small subset of the staff can understand it. And this is not because they have the high skills required to operate a very sophisticated tool. On the contrary, the tool itself is very basic. But you do require high skills to decipher the cryptic degenerative language in the E-mail being used to describe it.

Some may make the counterclaim that since some parts of the E-mail contains some code written by *Game Programmers*, the E-mail was not meant for everyone. It was only meant for highly skilled staff.

But, firstly, this has no bearing on the conclusion that the language of the E-mail is degenerative. The ones who make this counterclaim do so because they presumably know how to write code. And they recognise words in the E-mail which are code words from some programming language. However, that means that those who do not know how to write code, that is to say the vast majority of the staff involved in the production process, would not recognise these words. And the inclusion of these words in the language of the E-mail degenerates the language from a natural language to a pidgin language: one part natural language, one part programming language.

Secondly, the code in the E-mail was not the main subject of the E-mail. The main subject of the E-mail was how you format the headlines for the news articles that appear in a football season, during a game, using a tool. This tool is made up of three smaller tools called

1. Logan
2. text exporter
3. DevStrings

The 'Logan' tool is a Spreadsheet that all the staff can use to edit the text which formats the headlines. An omission which adds to the degeneracy of the language of this E-mail.

The 'text exporter' tool is a tool that converts text in the Spreadsheet into another form, a file or Database which is read by the game when it starts. All the staff can use this tool to manually perform the conversion at any time. In some parts of the E-mail, it is referred to as 'text exporter'. In other parts it is referred to as 'exporter tools'. An inconsistency which adds to the degeneracy of the language of this E-mail.

The 'DevStrings' tool is a *software library* which contains provisional text which formats the headlines that appears in the game. Alongside the text from the 'Logan' and the 'text exporter' tool. Until that provisional text is moved from 'DevStrings' to 'Logan'.

Now you may claim that the 'DevStrings' tool is only meant for *Game Programmers* and therefore the language for that tool was only for highly skilled staff. And therefore the E-mail was only for highly skilled staff.

But the other two tools in the E-mail were not meant only for *Game Programmers*. Those tools were meant for the rest of the staff as well. Yet the language in the

E-mail describing all three tools is the same degenerative language. This includes the cryptic words you see in the E-mail describing the news headlines e.g.

#3-Number# year old #1-Player# has suffered #2-Injury#

These are literally the words you see when you use the 'Logan' tool.

Thirdly, even the code from a programming language you see in that the E-mail is degenerative and incomplete. It is abbreviated with the word 'etc.' which does not come from any programming language. Therefore, even if you understand how to write code, you cannot completely understand this code. And the conclusion that the language in the E-mail describing the tool is degenerative still holds. Indeed the inclusion of this code supports that conclusion even further.

Such is the effect of the degenerative language of the *Software Evolution Process* that some of the staff miss many basic points about the *software production life cycle* and come to many misconceived conclusions. For example, they conclude that *private conversations* between staff are necessary when most of the time they are not. They conclude that the vision for the software user, the appearance of the *User Interface* and the components of the *Interface*, have nothing to do with the User Specification. Since they hear staff use words like 'the design vision', instead of the 'vision for the software user'. They hear staff use words like 'the quality of the art production', instead of the 'appearance of the *User Interface*'. They hear staff use words like 'the fun play mechanics', instead of 'the components of the *Interface*'. And they hear staff use words like '*game design*', instead of 'User Specification'.

They conclude that there is nothing wrong when they hear such phrases like

'We don't need to update that! The design vision and the quality of the art production, and the fun play mechanics have nothing to do with the game design'.

Nevertheless, as the quality of the language degenerates over time, in the *Software Evolution Process*, it affects the overall productivity of all the staff. And that in turn affects the overall success of the project. This has been confirmed by *research studies*,[32] which showed that improved communication contributed to more successful projects, increased new business and innovation, at the end of it.

The end of a *Software Evolution Process* is, however, not so promising. For it shares the same destiny as the *Tower of Babel*. The builders of the tower became confused and abandoned the project. Likewise, towards the end of a *Software Evolution Process*, chaos slowly begins to reign supreme. Some of the symptoms of this chaos are that many try to take the opportunity to market themselves. As the release date for the Computer Game approaches, they suggest changes either to the *game design, game modules, game-engine, Game data* or tools which market their ideas. They make no attempt to justify themselves. Instead they make negative propositions by asking,

Why can't you do this?
Why can't you do that?

But as many know it is difficult, if not impossible, to prove a negative. You have to search through all possible cases, to make sure that there is not one case where the negative proposition is false. In the context of a *Software Evolution Process*, you have to search through all cases, through all possible features, of the *game design* or *technical design*, for the one case where the negative proposition is false. That is to say, a proposed change to the *game design* has a negative effect on another feature. However since, as is typical at the beginning of a *Software Evolution Process*, these documents would not be complete, there would be no limit to how far you have to search. Thus, in effect, these opportunists do not prove themselves but place the onus on others to prove them wrong. And using this tactic, they cause traumatic changes to the *game design*. So much so that getting towards the end of the production process, many of the staff cannot wait for it to end and to see the back of the project.

Not only the opportunists amongst the staff of the *Software Developer* but those amongst the financial backers too take the opportunity to market themselves. They bring forth propositions to market their ideas, which end up putting added pressure on the production process.

For example, they may propose the premature release of photos of the game, to the press, to generate publicity. Even though these photos may contain features which either have only been partially implemented, not implemented at all or may not be in the final product.

On other occasions they may propose wholesale changes to the *User Interface*, one or two months before the release of the game. Even though these changes would be, at best, cosmetic and, at worst, endangering the quality of the final product.

On other occasions they may propose wholesale change of the theme of the game, all the while using the same *game modules*, *game-engine* and other tools as before, to develop it. Even though the game may have changed not only genres but art forms. It may have changed from being based on a film to being based on a sport; and from being based on a drama to being based on an athletics competition.

In the extreme case, the financial backers may personally intervene. And they may, for example, propose monitoring the day-to-day work of the *Software Developer* on their premises. So that, a few weeks before the date of its release, they could test the game on-site and increase the rate of feedback. And after the first version has been released, some *Software Developers* hire extra staff for the production of the next version, to reassure investors. But, in either case, this merely aggravates the chaos of the *Software Evolution Process*. For the greater the number of contingents participating in the *Software Evolution Process*, the greater the number of languages which develop from these contingents. And the more different languages develop, the more confusing the overall languages becomes. And the more confusing the overall language becomes, the less productive the staff become.

1.4   THE EFFECT ON CREDIBILITY

Hence, at the end of the *Software Evolution Process*, there is an incredible sense of relief, as the growing chaos finally comes to an abrupt end. Most of the tools that were developed by the staff, during the process, cannot practically be re-used. And privately, this technology would be quietly and completely overhauled before being used again, if at all. But publicly, many of the staff will claim to be satisfied by the technology they have produced during the production process.

At least, this is what they will say at the meeting which the *Software Developers* organise at the end of the project. The *Developers* call this meeting a *Post Mortem*.[33] Is the name merely irony, or do they know that the game is dead on arrival? This is one of many questions that those who take part in the meeting ask themselves.

Could they have performed better? Could the tools have performed better? Could the *Game Designers, Game Artists, Game Programmers, Sound Designers* or *Game Testers* have performed better? Was the incomplete *game design* responsible for the tumultuous experience which they had just gone through? Or was it all due to the pressure from the financial investors?

Some, with experience of previous projects, will be left with a feeling of Déjà vu which they cannot quite place.

Others will be left wandering if they had missed something. Was there a phase of the production process which they should have been involved in? Was there a decision, which they should have been privy to?

Although all the staff will sense that something is missing, they will not agree about what this is. And they will all be ambivalent about the role the *Software Evolution Process* itself played in this sense of loss. Instead, they will try to look to the components of the *Software Evolution Process* for an answer, the *game design*, the *technical design*, the tools and so on. And yet they will find no satisfaction.

But if they were to carefully look back to the beginning of the process, they would find the answer. They would see that the only thing amiss from the project was its credibility. For at the heart of the *Software Evolution Process* is the question of credibility.

As has already been explained, the *Software Evolution Process* is a default process. It is the minimal process that arises from the absence of any credible plan to build the Computer Game. There is no formal announcement that the *Software Evolution Process* has begun. Instead, there is simply either, at worst, no presentation of any plan or design for building the software. Or there is, at best, the presentation of half a plan, along with half a design, for the first couple of weeks or months of the project, in a meeting at the beginning of the process. Occasionally accompanying this may be euphemistic references to the *Software Evolution Process*, when someone ask questions about the details of the plan or design in the meeting. These include words like 'trial and error' or 'prototyping'. In response to these questions, someone might reply

> 'We do have a team of people here, roughly 80, working on the project. The idea was that we will get a green light by now, and it would be full steam ahead. But we haven't quite reached there yet.
>
> Until we get the green light we can't keep burning the money on 80 people. There is a fundamental belief that this team should develop a Zelda Style game. But we need to make sure that we are not burning all that cash on the project until it is green lit.
>
> So we will have a small group of people working on the pitch document. Another group of people working on the Prototype. And another group will be focusing on an all ages multiplayer game aspect.
>
> We already have a two page document for the family friendly aspect of the game. But we need to start prototyping on that as soon as possible...'

That would be the only sign, if any, that the *Software Evolution Process* had begun. Namely, the *lack of a plan*[34] at the beginning of the production process.

For soon afterwards, into the void left by the absence of any credible plan for building the game, at the beginning of the *Software Evolution Process*, will enter an incredible proposition. That is, a theoretical process of Biological Evolution, in nature, that occurs over millions and millions of years, to generate minute changes in animals to effect their survival, could be used to generate wholesale changes in software to effect its success. All within the period of 18 months or less, that it takes to produce a commercial Computer Game, with between 60 and 80 staff. And all for the sake of giving that project credibility; primarily to its financial backers and secondarily to its staff.

This form of *Neo-Darwinism*[35] has precedent in other industries besides Computer Games. Other industries too realise the credibility which the theory of Biological Evolution commands. And they too attempt to take advantage of it, by using it to cover up their *default processes*.[36] So that they can give those processes credibility. They all have different ways of making this seem plausible.

But in the Computer Games industry, the *Software Developer* ensures the credibility of this audacious stunt through one fact. That is, you cannot provide a good counter example, to the feasibility of the *Software Evolution Process*. You cannot provide a counter example either from its final product, or from any of its components. These include the documentation, the tools, the staff and other resources that help build the software. These also include the language the staff uses to communicate. The standards of the process are set so low that there is little requirement for it produce good records.

For as has already been previously stated, the *Software Evolution Process* is ad hoc. It only has two requirements. Firstly, that it slowly evolves and grows the software over time. And secondly that the basis of this evolution was feedback from the software user. Even the final product would only represent the feedback from the software user that happened to be around, when the project literally ran out of resources (i.e. time or money). The final product would not represent the set of requirements for the process.

In short, there is no requirement for the process to be scrutable. And therefore any *Post Mortem* at the end of the process is redundant. The guessing games the staff end up playing, by examining the final product, or the components of the process, are the inevitable consequences of this inscrutability. These games are a distraction from the strategic failure to conceive of a credible plan, at the onset of the project. These games lend the process a level of credibility which it never had.

1.5   THE USE OF NDAs TO GIVE THE PROCESS CREDIBILITY

There are other devices used to give the *Software Evolution Process* credibility. One of them are *Non-Disclosure Agreements*[37] or *NDAs*.

Frequently accompanying the *Software Evolution Process* is some form of *NDA*, which the *Software Developer* requires the staff or interviewees to abide by. And if the staff or any observers were looking for good counter examples from the process, these *Agreements* would be amongst the best. For these would serve as perfect illustrations of the strategic flaws of the process.

These *Agreements* are practically unenforceable due to the inscrutability of the *Software Evolution Process*. Short of a signed confession, the best that a *Software Developer* could hope for would be circumstantial evidence of violations. The *Developer* could only show that an alleged original feature, in the final product or earlier versions, existed prior to the one found in a competing product. And that former staff of the *Developer*, involved in the *Software Evolution Process*, joined a competitor prior to the appearance of these features in their competing product. Otherwise, there is no way the *Software Developer* could show the genesis of any original feature, and how the accused were involved, to support their case. And the *Software Developer* could only dream of proving a case, involving technical innovations, taken from a project, and incorporated by a competitor into their *technical design*. For such innovations, and the genesis of these innovations, would simply not be documented.

Despite the existence of such *Agreements*, very few of the games produced by *Software Developers*, out of the thousands released each year, have been *original games*.[38] Nor indeed have these games warranted such *Agreements* in the first place. For the continuous evolution of the software, based on feedback from the software user, means that ideas very quickly become out of date anyway. Thus, any ideas which a competitor could discern from a *Software Evolution Process*, at any given point in time, would be misleading.

Any ideas which they discerned at the beginning, or the middle of the process, would be just as misleading to them as to the staff involved. For the way the process keeps mutating, slowly destroying any definitions it acquires, at the beginning of the project, also leads to the destruction of any ideas introduced during its course. And this produces a paradox.

On the one hand, the course of the *Software Evolution Process* is unpredictable, due to its constant loss of definitions. Yet, on the other hand, the final product contains little or no innovation. Since only ideas introduced towards the end of the process maintain some definition. And these invariably are unoriginal ideas, copied hastily from other competing products, into the final product, as time runs out, and the project heads towards a crisis.

But any of these belated ideas would be useless to a competitor. Since they could not hope to incorporate these into their game and release it before the *Software Developer* did. It would be contrary, to the evolutionary principles of the *Software Evolution Process*, for the construction of any part of the product to be documented, up to three months in advance. And no part of the game would conceivably be documented up to a year in advance, before being implemented. Thus giving any competitor, who got hold of that document, time to beat the *Software Developer* to the market.

Once you realise that the *Non-Disclosure Agreements* are unenforceable and unwarranted because they do not protect any original ideas, then you realise that these must serve some other purpose. And once you realise that the *Software Evolution Process* enters where there is no credible plan to build the software, in this case a Computer Game, you realise exactly what that purpose is. Namely, through *Non-Disclosure Agreements*, and other devices, the *Software Developer* gives the *Software Evolution Process* an aura of credibility. These devices act as a shroud that covers the *Software Evolution Process* in mystery.

These devices include the *game design*, the *technical design* and other documents. Each will be labelled with notices reminding the staff that it was 'Strictly confidential' (see Figure 1.3).

Even though each of the documents, the *game design* and the *technical design*, will largely be empty. But for a wish list of vague points which the authors hoped the process will pass by at sometime in the future. Even though all of these documents would already be covered by the *Non-Disclosure Agreements* which the staff signed.

All for the sake of exaggerating the secrecy and mystery surrounding the *Software Evolution Process*. So that, through this mystery, it achieves a certain level of mystique. And, through this mystique, it achieves a veneer of credibility, at its onset.

1.6   THE USE OF SECRECY AND MYSTERY TO GIVE THE PROCESS CREDIBILITY

Another device used to give the *Software Evolution Process* credibility which is closely related to *NDAs* is the use of code names.

Often *Software Developers* will give code names to their projects. To disguise the contents of the projects they were working on for a client. And the staff will be required to use these code names when discussing projects internally or externally with clients. These code names will be based on some kind of theme like colour e.g.

Project Red
Project Blue
Project Green

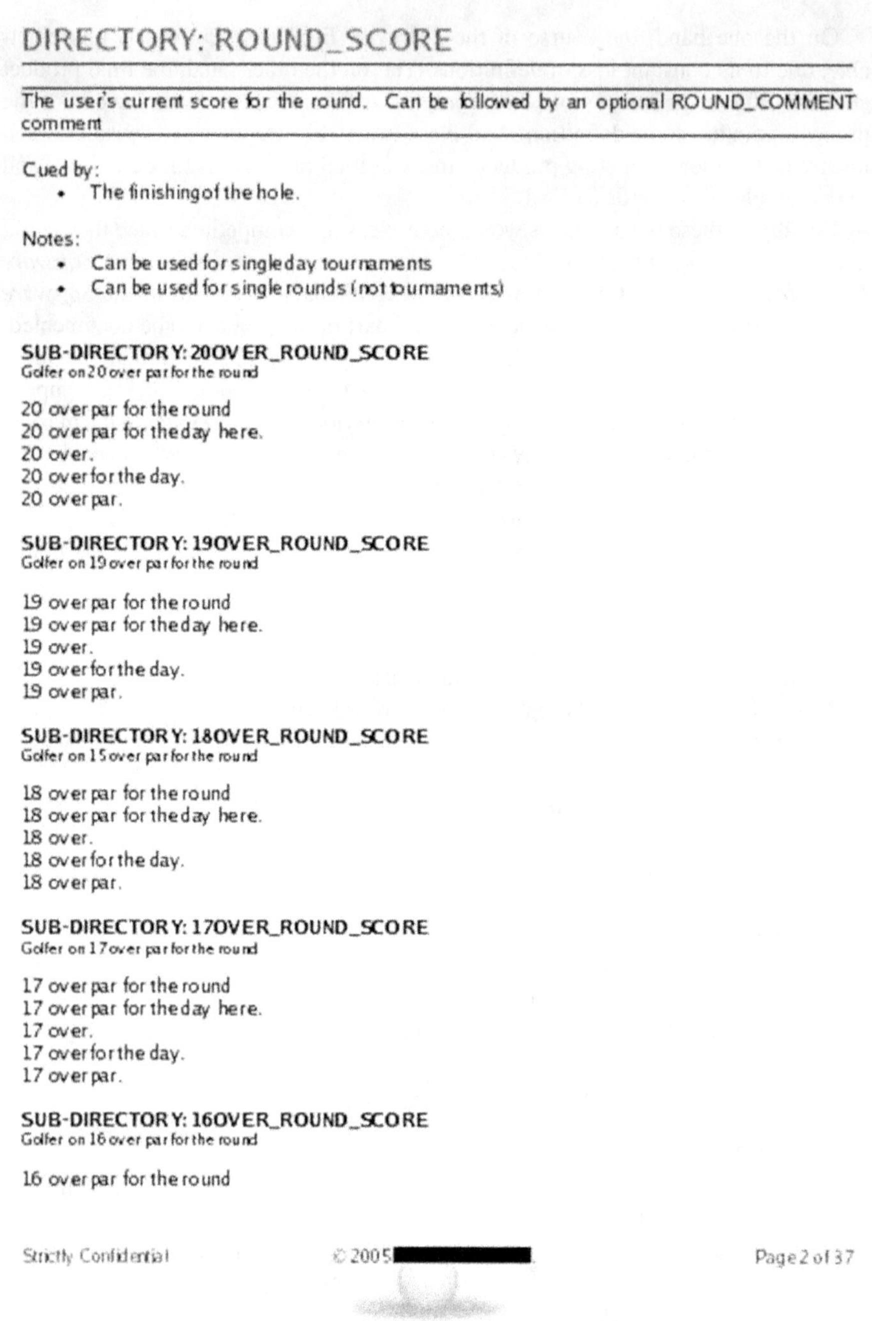

DIRECTORY: ROUND_SCORE

The user's current score for the round. Can be followed by an optional ROUND_COMMENT comment

Cued by:
- The finishing of the hole.

Notes:
- Can be used for single day tournaments
- Can be used for single rounds (not tournaments)

SUB-DIRECTORY: 20OVER_ROUND_SCORE
Golfer on 20 over par for the round

20 over par for the round
20 over par for the day here.
20 over.
20 over for the day.
20 over par.

SUB-DIRECTORY: 19OVER_ROUND_SCORE
Golfer on 19 over par for the round

19 over par for the round
19 over par for the day here.
19 over.
19 over for the day.
19 over par.

SUB-DIRECTORY: 18OVER_ROUND_SCORE
Golfer on 15 over par for the round

18 over par for the round
18 over par for the day here.
18 over.
18 over for the day.
18 over par.

SUB-DIRECTORY: 17OVER_ROUND_SCORE
Golfer on 17 over par for the round

17 over par for the round
17 over par for the day here.
17 over.
17 over for the day.
17 over par.

SUB-DIRECTORY: 16OVER_ROUND_SCORE
Golfer on 16 over par for the round

16 over par for the round

FIGURE 1.3   A typical example of a page from an incomplete game design in a Software Evolution Process of Slippery Games Inc. (Source: A typical game design from a Software Evolution Process of Slippery Games Inc. showing the excessive use of confidentiality to shroud in mystery unoriginal game designs. Anonymous. 2005.)

Or they may have random titles based on names of Greek Gods, precious metals or sauces e.g.

Project Zeus
Project Sapphire
Project Tomato

All of these names, like the *NDAs*, serve to shroud each project in mystery and secrecy. So much so that staff working at the same *Software Developer* but on two different projects cannot discuss what they are working with their colleagues.

Even though the staff may share the same technology and tools on the two different projects. And they cannot really understand these tools or technologies unless they understand the projects which these came from.

Even though the staff working on one project may sometimes find the code name of other projects in the files they use in their project. And they will be required to edit these files and substitute the code names of the other projects with more code names.

Even though sometimes the code name of these projects end up in the names of files, *game modules*, *game-engines* or *software libraries* used to build the final product.

Even though by introducing these code names in a project, the *Software Developer* will be deliberately introducing words which were meant to obfuscate what a project was about. And thus accelerating the degeneration of the language of the *Software Evolution Process*. That would otherwise naturally degenerate anyway. Due to the way the staff from different backgrounds will add words to the language based on their background. In the absence of any clear plan or design at the start of the process, which gives them a suitable language.

Even though by introducing these code names in a project, the *Software Developer* will be giving it a false name. And thus aggravating a project already suffering from a lack of identity and a lack of vision. Due to the way the *Software Evolution Process* begins with an incomplete *game design*.

Again, just like *NDAs*, these code names will be all for the sake of exaggerating the secrecy and mystery surrounding the *Software Evolution Process* of each project. So that, through this mystery, it achieves a certain level of mystique. And, through this mystique, it achieves a veneer of credibility.

1.7 THE USE OF INSCRUTABILITY TO GIVE THE PROCESS CREDIBILITY

As previously mentioned, there is no requirement in the *Software Evolution Process* to keep good records. And the process would begin with a lot of documents (e.g. the *game design* and the *technical design*, the User Manual, the test plan) which were either incomplete. Or these did not exist at all. Or these were subsequently not kept up-to-date after the process began.

As a result when errors surfaced later on, it would be next to impossible to investigate the source of the errors, since the process would be inscrutable. It would be next to impossible to tell when or why errors occurred because of changes made to the *game design*. It would be next to impossible to tell when the introduction of one feature conflicted with another introduced later on. It would be next to impossible to tell when the news headlines congratulating a new football manager for winning the first football match, conflicted with news articles with sarcastic headlines for the football manager winning the football season curtain raiser.

Some may say well you can tell when the changes to the files used to build the first feature were made and submitted to a Software Repository. And when you can tell when the changes to the files used to build the second feature was made and submitted to a Software Repository. And therefore you can tell when the error was introduced.

But as already explained, in the *Software Evolution Process* you may have two groups independently developing two features in the game, in two concurrent over-lapping sub-processes. And it later emerges that there is a conflict between the two features. And neither the times that the files used to build the first feature, nor the times that the files used to build the second feature, were checked into the Repository will tell you when or why the error occurred. The error occurred somewhere in between when both groups started developing the two features, and when they suc-cessfully built and tested those two features. Before they submitted the files to the Software Repository. And that time will not be documented.

Likewise, it would be next to impossible to tell when errors occurred because one feature was missing or not well-described. It would be next to impossible to tell when the player could not complete a quest in an adventure game because some item they needed to complete the quest was missing. When some special weapon or armour they needed to retrieve from one character in the *Game World*,[39] and give to another, had not been made. Or when the look, appearance and animation of the first charac-ter and second character had not been decided.

Some may say well you can tell when the quest was written into the *game design*. And you can tell when the quest was added to the *Game World*. And therefore you can tell when the error was introduced.

But those times do not tell you when the error was introduced. At the beginning of a *Software Evolution Process*, the *game design* will be incomplete, and some quests in the *game design* will also be incomplete. It is not clear at that point that any errors have been introduced. So long as before the quest was tested in the *Game World*, all of the items, locations, characters and animations that were required for the quest had been added to the *Game World*, then there would be no errors. But the length of time required to complete building these items, locations, characters and animations in the *Game World* would be indeterminate. That length of time would not be documented. And therefore you could not tell when the errors with the quest were introduced.

Likewise, it would be next to impossible to tell when errors occurred due to changes introduced in the *technical design* for a Formula 1 Racing game. When the sound generated by the engines of the cars after a crash began to sound very similar to the sound generated by the engines of the cars revving up on the start grid.

Some may say that well you can tell when the files used to generate the sound of the engines of the cars were changed. And therefore you could tell when the sound generated by the engines after a crash was introduced.

But did the error occur when the sound of the engine of cars revving up on the start grid was generated? Or did the error occur when the sound of the engine of the cars after a crash was generated? If the *Software Evolution Process* began with an incomplete *game design*, which did not specify how the engines should sound on the start grid or after a crash, then how do you know when the sound generated on the start grid or after a crash was in error? And if you do not know what the engine should sound like, then how can you tell when the error was introduced?

All of these errors in the *Software Evolution Process* will be explained away as being due to the inscrutability of the process. And in the absence of any conclusive proof as to the cause, a lot of provisions would be added to the process to avoid the errors in the future.

A special *Game Programmer* would be assigned to generating the sounds of the engine of the cars for the racing game. To make sure the sounds it generated were unique. A special *Game Tester* would be assigned the task of checking all the quest could be completed in the adventure game. A special *Game Designer* would be assigned the task of examining the news articles in the football management game. To make sure there were no conflicts.

All of these makeshift assignments will be used to patch up the *Software Evolution Process*, and give the process credibility so that it could continue.

1.8 THE MYTHICAL MAN MONTH

As mentioned in the previous subchapter often a special member of staff is designated or assigned to patch some of the problems that emerge from the *Software Evolution Process*. This leads on to another device which is used to give the *Software Evolution Process* credibility.

And this is a myth known as the *Mythical Man Month*.[40] The myth is that by doubling the staff involved in the *Software Evolution Process*, you can halve the time it takes to produce a game. It is the *Mythical Man Month* that gave the *Software Evolution Process* its name. It was the attempt to come up with a formal theory to explain why doubling the staff involved in a *Software Evolution Process* did not half the time it took to complete a project. That led the man who attempted to formalise this theory to coin the phrase Software Evolution to explain the phenomenon.

But so far he has not been able to come up with a theory which satisfactorily explains the phenomenon. And stops the *Software Evolution Process* becoming more chaotic and unpredictable when you double the staff involved.

Since you have greater communication problems in the *Software Evolution Process* when you double the number of staff involved. And you need more time to train the staff you added to the process. To get them to understand the language of the process. You need more time to get them up to speed before they can begin to contribute.

Nevertheless, many *Software Developers* still continue to believe in the myth. And that by doubling the number of their staff involved at the end of one process, and the beginning of the next process, they can reduce the time it takes to make the product in the second process. And they can increase the rate of feedback in the *Software Evolution Process* and make it more stable.

Or so they will say to their financial backers, who query the process, after a game has been released, which either had lots of errors or was delayed. They will reassure the financial backers by reiterating the *Mythical Man Month*. That by doubling the number of staff, and increasing the rate of feedback, they will stabilise the *Software Evolution Process*. All to give the process credibility.

1.9   THE USE OF THE PROMISE OF RAPID FEEDBACK TO GIVE THE PROCESS CREDIBILITY

As mentioned in the previous subchapter, the *Mythical Man Month* is one of the devices used to give the *Software Evolution Process* credibility. Part of this myth is the promise of rapid feedback can make the process more stable, less chaotic and more predictable.

In theory, if the errors that emerged in the production process could be caught earlier on, then these could be addressed earlier. And stop them growing into major problems later on. In the Computer Games industry, the more *Game Testers* or people you have performing a similar role you have in the process, to test the game, the more eyes you will have examining the product. And the more eyes you have examining the product, the more likely it will be you will catch errors at an earlier stage.

In practice, this does not work. The *Game Testers* do not have anything to guide them when performing these tests. There is no standard to measure the game by. In the *classic software production life cycle*, you would have a test plan that would be your standard. But in the *Software Evolution Process*, they do not have a test plan or even a User Manual to guide them in the test. Since, in the *classic software production life cycle*, a complete test plan requires a complete User Manual which is used to produce that test plan. And a complete User Manual requires a complete User Specification or *game design* which is used to produce the User Manual. But, in a *Software Evolution Process,* you will not have a complete *game design*.

Therefore the *Game Testers* have to use their own intuition when conducting these tests. They have to use their own subjective judgement when playing the game, to decide when something looks right or wrong. The quality of the tests they conduct varies greatly because their experiences, which informs their intuition, varies greatly. And furthermore, there is no standard (i.e. a test plan) to measure the game by. There is nothing to stop them from testing the same areas of the *Game World*, the same features in the *World*, the same menus or the same commands in the *User Interface*. As a result lot there is a lot of redundancy in the test they perform.

To compensate for the lack of a test plan, the *Software Developers* attempt to add more Game Testers to give the Software Evolution Process credibility, to make it more stable. But the more *Game Testers* you add to the *Software Evolution Process*, the greater the redundancy they will produce. And the greater the redundancy they produce, the more resources (i.e. staff or time) would be required to go through all the tests and eliminate the redundancy.

And to compensate for the lack of a User Manual, the *Software Developers* attempt to use automated tutorials built into the game. That introduce new players to the game and show them how to use the *User Interface.* All for the sake of giving the process credibility. When it becomes self-evident, to the *Game Testers*, towards the end of the process, that a lot of changes had been made to the *User Interface* since the beginning. They were not aware of most of the changes. And there had been no document that kept track of all these changes that could explain it to the players. So some device was required to explain it all.

However, the tutorials introduced into the game to explain it all, at the end of the process, end up making software more complex at a time you want to reduce complexity. Now the *Software Developer* has to add more parts to the software to build these tutorials. They will require more work from the *Game Artists, Sound Designers, Game Designers* and *Game Programmers.* All of that work would be unplanned.

And this new work will add more relationships, more dependencies between these new parts of the game and other older parts of the game. Now if the *Software Developer* changes the *User Interface* and adds new components, new buttons, commands, then they will not only have to edit the old parts of the game to use the new components. They will also have to edit the new parts which teach the players how to use the new components.

And when the *Game Testers* tests these new components of the *User Interface*, they will have to test them both in the main game and in the tutorials. And the greater complexity introduced by these new parts, the greater number of dependencies between the old and new parts, and the testing of these new parts, will introduce new errors. That the *Software Developer* would not otherwise have if they had a User Manual that was separate from the game, which explained the *User Interface.*

1.10   THE DECLINE AND FALL OF CREDIBILITY

All the devices which are used to give the *Software Evolution Process* credibility only help it to achieve this credibility by implication. And, it achieves this only for a limited time. Since the devices which give it this credibility are all ineffectual.

Eventually, all of these devices end up undermining its credibility. These included

1. *NDAs,*
2. excessive secrecy and mystery surrounding projects,
3. code names for projects,
4. the *Mythical Man Month,*
5. the promise of rapid feedback.

This credibility is lost, first and foremost, on the staff who follow the *Software Evolution Process.* They subsequently become aware that the game they are producing contains little or no innovation to warrant such *Non-Disclosure Agreements* and the excessive secrecy and mysterious aura which surrounds the project.

Likewise, the code names given to the projects only end up obfuscating the vision for the project. And undermine a project already suffering from a crisis of identity because of the lack of a clear plan or complete *game design* to build the game.

Similarly, the inscrutability of the process, which helped maintain any credibility it achieved, with the financial backers, at the beginning, ends up undermining it. For the leadership of the project, and its financial backers, subsequently become aware of the consequences, of the lack of written designs or plans for building the software. They become aware that the leadership cannot account for some of the errors that arise later on, because of this inscrutability.

Likewise, the *Mythical Man Month* ends up undermining the *Software Evolution Process*. Any additional staff introduced into the process late on, by the financial backers or the *Software Developers*, to restore its credibility, ends up undermining it. For the greater number of staff requires greater communication between the staff. And it requires more time for the new staff to learn the language of the process, and get up to speed, before they can start contributing. The older staff, who have had previous experiences of the *Software Evolution Process*, realise that the quality of the work they produce starts degenerating. And it becomes far worse than when they were working with fewer staff or alone. Even though they may not recognise this as a classic symptom of the *Software Evolution Process*. Namely, as has already been explained, the more contingents participating in the process, the more chaotic production becomes.

Correspondingly, the continuous changes to the *game design* and evolution of the game, based on rapid feedback from the software user, which gives the process some credibility at the beginning, end up undermining it. The enthusiasm of the staff, at the beginning, gives way to cynicism as the changes become exhausting and seemingly never ending.

Finally, towards the end of the process, opportunists take advantage of the steady decline, in the credibility of the project, to market their own ideas to salvage this credibility. They propose changes to the *game design* which cannot be refuted, by making negative propositions

Why can't you do this?
Why can't you do that?

These propositions cannot be refuted for exactly the same reason that the feasibility of the *Software Evolution Process* could not be refuted at the beginning of the project. Namely, there was no requirement for the *Software Evolution Process* to keep good records or documentation. And therefore the process would be unscrutable. The *game design* and *technical design* would be incomplete and not kept up-to-date. And it may be possible that, in some future change in the *game design* or *technical design*, these negative propositions may become false. And some proposed change in the *game design* or *technical design* may become feasible.

The credibility of the entire process reaches its peak at the very beginning. And this is also, unsurprisingly, the high point of its documentation. From this apex, the credibility of the process rolls downwards to its lowest point, at the end of the project, just before the release of the Computer Game. And those products whose processes lose all credibility, before this date, simply never see the light of day.

1.11   THE POST MORTEM

Despite the fact that all the devices used to give the *Software Evolution Process* credibility inevitably fail, successive generations of *Software Developers* continue to lend the *Software Evolution Process* credibility. The *Mythical Man Month*, and the *Software Evolution Process* from which it stems, have both become pervasive. So much so that, in the *Post Mortems* at the end of the process, in response to the staff that were sceptical of the process, the *Software Developers* who advocate it expect the sceptics to provide examples that illustrate its flaws.

By which they mean they want the staff to provide counter examples from inside the phases of the production process itself i.e. *Pre-production, Production and QA*. They do not want counter examples from outside of these phases. That show that the *Software Evolution Process* can never achieve what were supposed to be its two main characteristics. Namely, they have no interest in counter examples which show that it cannot *evolve software without degenerating*, nor can it depend on *feedback from the software users*.

Instead, they presume to conduct the *Post Mortem* on the basis of a fallacy known as begging the question. What could have been done better during the process? This question assumes that the *Software Evolution Process* has ended, the staff can scrutinise the results and they can assess what could be improved in it. The question invites the staff to assume the conclusions of a *Post Mortem* i.e.

1. the process has ended
2. the process can be scrutinised
3. the process can be improved

Without any proof to support those conclusions. All of these conclusions are false.

Firstly, the conclusion that the process has ended is false. As previously explained, a *Software Evolution Process* is an *open-ended process* which never ends. A real *Post Mortem* could not reach the conclusion that the process has ended. There will be no document which clearly states, this or that is the end-point. This or that should be in the final game. So that you could use that document to conclude the process had ended.

On the contrary, there will sometimes be documents which prove the opposite case, that the process has not ended, such as the number of *Bugs*[41] being reported in

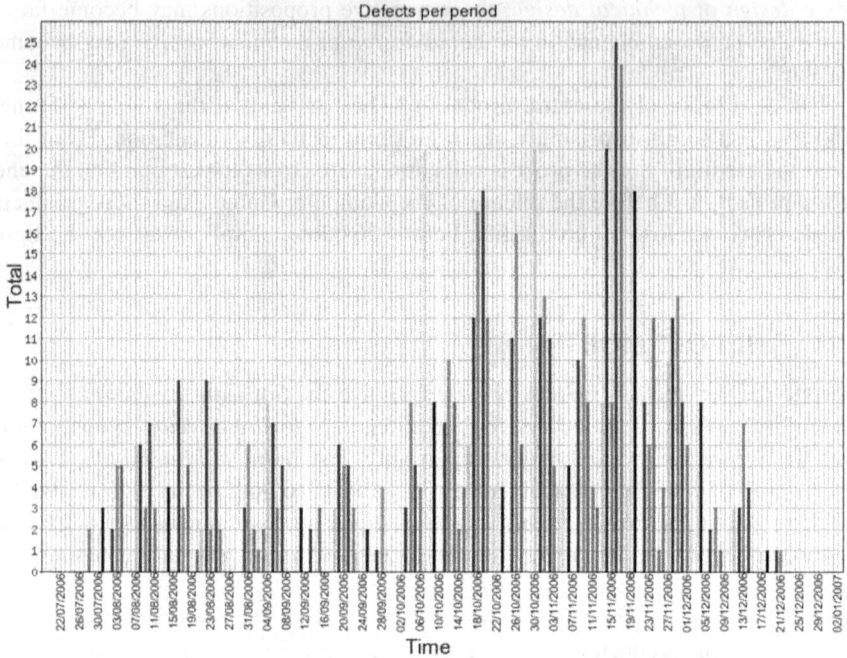

FIGURE 1.4   A typical example of a bar graph showing the new errors reported in a Bug Database, each day, over the course of six months, during a Software Evolution Process of Slippery Games Inc. (Source: A typical report of new errors in a Bug Database, during six months of the final, testing phase of a Software Evolution Process of Slippery Games Inc. The game was a football management game. It should have been released before the beginning of the football season in August. But you can see Bugs being reported in the game all the way up to December.)

the *Bug Database*.[42] This will show that there were still be some outstanding *Bugs* left when the process nominally ended, and game was released (see Figures 1.4 and 1.5).

Secondly, the conclusion that the *Software Evolution Process* is scrutable is also false. As previously explained, the *Software Evolution Process* also has no requirement to keep good records. The *game design* will be incomplete. The *technical design* will be incomplete. Therefore a real *Post Mortem* could not reach the conclusion that the process could be scrutinised.

Thirdly, the conclusion that the *Software Evolution Process* could be improved is false. Since the process would not have the records required to allow the staff to identify the areas in it that could be improved. A real *Post Mortem* could not reach any conclusions about how the process could be improved.

This then begs the question! If a real *Post Mortem* could not possibly reach these three conclusions, then what would be the point of the *Post Mortems* at the end of a *Software Evolution Process*? What purpose would they serve?

The answer to that question is simple. The *Post Mortem* meetings merely serve as a release valve, as a form of therapy. To allow the staff to release their frustrations that naturally build up during the *Software Evolution Process* due its major flaws. The

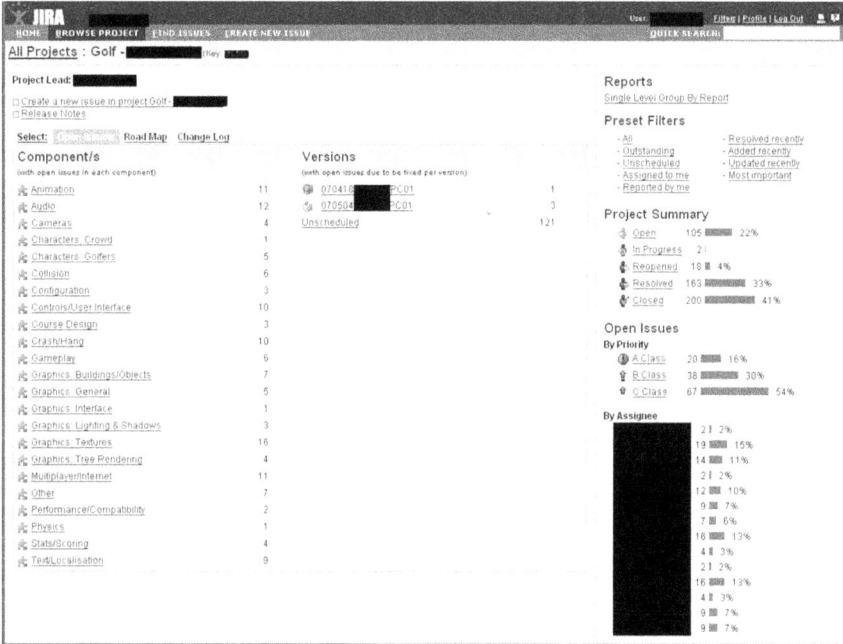

FIGURE 1.5   A typical example of a Web Page showing the high number of Bugs reported in a Bug Database during the QA phase of a Software Evolution Process of Slipper Games Inc. (Source: A typical report of errors in a Bug Database, during the final, testing phase of a Software Evolution Process of Slippery Games Inc. Anonymous. June 2007.)

most important of these will be the communication between the staff, the way their language degenerates with time and the way the software degenerates as it evolves.

'Please add any comments related to the working with other teams here:

"You are awesome!!"

Publishing reshuffling the priorities of bugs to suit their needs is unhelp-ful i.e. "They make their personal issues top priority so you end up fixing a wrong 'cuddle a teddy bear' animation, rather than game breakers."

Design should be more assertive and not allow Publishing and execs to change designs.

"[Last minute changes were reasonable...] if they weren't camouflaged as a feature request."

"xbox save system. This was a last minute bug request, which actually turned out to be a whole rework of it."

"...kudos to our compliance teams ([Redacted] and others). They have been very helpful in answering questions."

R---- Tech sending bugs straight back, requesting that we verify the bugs are actually for them, when that should really be their job.

"Everyone working on [Redacted] is generally quite nice so no complaints here!"

"Being part of [Redacted]Systems I sometimes find myself not being informed of being part of discussions about features directly involving our systems."

"Sometimes g--- tech don't like us sending things their way and we have to fix their bugs."

"Members from other teams who chipped in in the run up to [Redacted] were spot on."

"Production have been spot on in almost all interactions."

"Additional design or dev bugs coming in from publishing are sometimes really poorly scoped or have insufficient information."

"More fleshed out, more timely designs for the tasks would be appreciated."

"Last minute change requests have been a little extreme in a few cases."

"G – tech were good!"

....

1. Were you happy with the kinds of tasks that were assigned to you?

"Yes!"

"Can't complain"

"Yes. My work was pretty much always scoped to my expertise."

"Yes, most of the time, but sometimes bugs like to hide or be difficult to track."

"Most of the time I was bug fixing monkey, but retrospectively it was ok... but frustrating"

"Yes, happy to take on more"

"I'm happy with any task, I like challenges."

"Yeah for the most part.'"

Source: A typical Post Mortem Document from Slippery
Games Inc. 2017. Redacted. Anonymous

1. *Overall, how do you feel [Redacted] went?*
   It could have been managed better. It was initially going to be released around September? But then it was moved back to December? And then it was moved again back to March? And then it was moved back again to June? Missing so many deadlines seems bad for the reputation of the company.

2. *Which aspects went well during the development of [Redacted]?*
   The optimisation which some how managed to improve the performance of [Redacted] to the point where the Producers ([Redacted[?) were satisfied with it. The User Interface seemed to go through a lot of changes and iterations but somehow the people working on it managed to keep up with the changes.

3. *What expectations did you have about working on [Redacted], that did or didn't happen?*
   I was expecting [Redacted] to be released in December. But that did not happen. I expected to rely on other people's expertise to help me understand the project. And they did help when they could especially ....

4. *What frustrations did you encounter working on [Redacted]?*
There was little sense of a Game Design, and how far the production process was to completing that Game Design. It seemed that everything was
in a state of flux, features were being added or removed, without any
announcement.
5. *What one thing would be most important to improve for our future projects?*
This can be something you have already mentioned that you would like to call out as the most important thing to you
There should have been more show-and-tell, to give people an idea of
what other people were working on and how far they had progressed.
6. *Anything Else?*
There was little or no documentation of work done. A lot of communication seemed to rely on informal verbal communications, when other people may be busy and do not have the time to stop their work and explain.
And this is also a flaw in the [Redacted] Engine. This Post-Mortem is far
too late to feedback into other Projects which are about to finish or halfway through.'

Source: A typical Post Mortem Document of Slippery
Games Inc. 2020. Redacted. Anonymous

The *Software Developer* always concludes the *Post-mortem meeting* or therapy session by asserting that there was no ideal process. To placate the staff's frustrations and get them to accept that it is only natural. Although there are several ideals but not one that the *Software Developer* would be willing to admit to. Most *Software Developers* do not even know what the *classic software production life cycle* is, especially in the Computer Games industry. And those that do reject it as the ideal.

Most *Software Developers* do not even know where the *Software Evolution Process* came from, and that the man who gave it its name has written several laws describing it. Those that do know reject these laws as the ideal.

Instead, some have chosen to hold up the theory of Biological Evolution as the ideal. But as has already been explained, they are at best being credulous and at worst fraudulent. Biological Evolution is generative and never ends, where as the *Software Evolution Process* is degenerative. Although, theoretically, its phases are infinite, practically the process always, rapidly and effectively grinds to a halt, shortly after it begins. The vast majority of the modifications to the Computer Game, after the first rounds of designs have been written and built, are remedial. Thus the software is a spent force long before the day of its release.

Besides, who would, in all sincerity, choose to be part of a software production process which never ends? What would be the chances? That one of the longest, most complex processes in nature, would turn out to be an exemplary model for an ad hoc process? A default process which was merely incidental to a project suffering from a lack of credibility?

NOTES

1. *Classic software production life cycle.* A production process that follows the analysis, design, implementation, testing, installation, maintenance and retirement of software. See Glossary.
2. *Software design.* A breakdown of the software components, tools and techniques that will be used to build and assemble software that meets a User Specification i.e. a customer's requirements. A breakdown of the software procedures and data that will be used to build a software module.
3. *Software module.* A small piece of software, which is a component part of a larger computer program. Each solves one facet of the overall problem the programme was designed for.
4. *Software data.* Information suitable for computer processing. *Game data.* General information which is shared between software modules e.g. the name of items in the world, commonly used text, 2D images, 3D models or sound. *Abstract data.* Special information which is designed to be used by a single software module.
5. *Software library.* A collection of computer programs, or software modules, which perform commonly repeated task on computer hardware e.g. reading data from, and writing to, computer files.
6. *Game design.* A term sometimes used to refer to the User Specification of software in the Computer Games industry. It is a description of the goal of a game, the different stages, the progression and the User Interface through these stages.
7. *Interface.* A common boundary. *User Interface.* The set of components (e.g. images, messages, commands or menu options) that allows a user to interact with software. *Programming Interface.* The set of components (e.g. procedures or data) that allows one software module to interact with another.
8. *Technical design.* A term sometimes used to refer to a software design, in the Computer Games industry. It is a description of the software modules, data, tools and techniques that will be used to implement a game design.
9. *Game module.* A software module which is used to implement unique aspects of a game.
10. *Game-engine.* A set of software modules (or library) which were designed to be re-used to make different games.
11. *User Manual.* An instruction booklet, for software users, which explains how to solve a problem using the software.
12. *Pre-production, Production and QA.* The differences between the names of the phases, of the production of Computer Games, and other software, stem from the crisis of identity which the Computer Games industry suffers from. See Glossary.
13. *Software Developer.* A person or company that produces software.
14. *Platform.* A marketing term for a computer hardware or Operating System or third-party game-engine that a Computer Game can be built and sold on.
15. *Game-editors.* One or more tools that allow the elements of a game to be edited. These elements may either be menus, locations, characters or other items that appear in the game. See Glossary.
16. *Software Evolution Process.* A name given to any of a large set of ad hoc, non-linear, software production processes that are meant to evolve a piece of software to meet a software user's requirements. These are based on rapid feedback from the user. See Glossary.
17. *Practical application (of Biological Evolution).* Advocates of the application of the theory of Biological Evolution, to software production, can be found in many quarters of the Software industry. See Glossary.
18. *Open-ended process (of Biological Evolution).* The extinction of a species is not the end-point or goal of Biological Evolution. The goal is survival; survival of the fittest. See Glossary.

19. ***Art Pipeline.*** A manual, distinct sub-process of the production of Computer Games, for generating artwork. See Glossary.
20. ***Build Pipeline.*** An automated, distinct sub-process of the production of Computer Games, for periodically building and testing the game to ensure no errors have been introduced into the building process, due to rapid changes. See Glossary.
21. ***Evolve software without degenerating.*** M. M. Lehman, who gave the Software Evolution Process its name, has always found the process degenerative and denied any link between it and the generative Biological Evolution. See Glossary.
22. ***Feedback from the software users.*** Virtually all commercial software licences exclude the software users from the production process. See Glossary.
23. ***Game Producer.*** An employee of a games company in charge of the overall production of a game, from getting its financing, through its analysis, design, implementation, testing, to its release.
24. ***Game Designer.*** An employee of a games company responsible for game designs.
25. ***Game Tester.*** An employee of a games company who tests the software at the end of its production.
26. ***Game Programmer.*** An employee of a games company who writes the software for games.
27. ***Game Artist.*** An employee of a games company who makes 3D models, 2D images and animations used in a game.
28. ***Sound Designer.*** An employee of a games company who creates and records the sound and music played in a game.
29. ***Tower of Babel.*** In his book, 'The Mythical Man Month', Frederick P. Brooks compared the confusing language that arises in the Software Evolution Process to the building of the Tower of Babel. See Glossary.
30. ***Private conversations.*** Some isolated observers may view the private conversations, and the decisions which come out of these, as beneficial to the productivity of a Software Evolution Process. But that would be a mistake. See Glossary.
31. ***Memorandum.*** The primary source of the explanation of the tools used, in a Software Evolution Process, are memoranda. See Glossary.
32. ***Research studies.*** Studies conducted for the International Business Machines Corporation (IBM) and the Microelectronics and Computer Technology Corporation (MCC) showed that improved communication had a beneficial effect on productivity. See Glossary.
33. ***Post Mortem meeting.*** A meeting conducted at the end of a software production process, by the staff involved, to retrospectively examine the pros and cons. And to decide the lessons to be learnt from the experience. See Glossary.
34. ***Lack of a plan (in Software Evolution).*** Instead of the plan (i.e. the game design and technical design) providing for contingencies, the plan itself becomes a contingency, i.e. a future event which cannot be predicted with certainty. See Glossary.
35. ***Neo-Darwinism.*** The synthesis of a modern theory with the theory of Biological Evolution by Charles Darwin. Usually this refers to the synthesis of genetics with Darwin's theory. See Glossary.
36. ***Default processes (based on Neo-Darwinism).*** Other industries tend not to make explicit references to the theory of Biological Evolution, through the names of their default processes. They are more pragmatic about what they call these processes. See Glossary.
37. ***Non-Disclosure Agreement.*** A confidentiality agreement not to divulge information relating to a software project, to anyone outside that project. The agreement would normally be used to stop the disclosure of original inventions. And it may sometimes be incorporated into the contracts of the staff involved in that project. But the agreement has often been abused. See Glossary.
38. ***Original games (released each year).*** The majority of the Computer Games that have been released each year have tended to be clones or sequels of successful games from the past. See Glossary.

39. ***Game World.*** An imaginary world space in which a game takes place.
40. ***Mythical Man Month.*** The theory that if it took one person a certain amount of time to complete a task, it would take two approximately half the time to complete that same task. See Glossary.
41. ***Bug.*** A software error. The name comes from an anecdotal story about an error caused by a moth short-circuiting an old computer.
42. ***Bug Database.*** A Database of the errors found in a software product. See Glossary.

2 The Solution

Clearly, the solution to the problem of building a game, which lacked definition, would be not to default on the *Software Evolution Process*. This would merely undermine the credibility of the production process. Instead, in its place, the *Software Developer* should substitute a process which had credibility. A process would not have to be perfect to be credible: merely open and honest. A credible process would give at least the language, the staff used to communicate, definition. And this would make them more productive, than in a *Software Evolution Process*.

One obvious way this definition could be provided would be for the *Software Developer* to complete, or attempt to complete the *game design*, at the beginning of production. But, as has already been mentioned, it may not be in the *Developer's* interest to write a complete *game design*. If the *Developer* seeks financial backers to publish the game, the financial backers may request changes to some, if not all of the *game design*. And the *Software Developer* may waste a lot of resources trying to meet these requests. They may have to edit a lot of the *game design*, the *technical design*, the tools and the game they had already built.

It may be, in the best case, that either the time it could take to edit the game in this manner could be negligible. Or that the *Software Developer* could build the game first, before seeking a financial backer. And so they could be in charge of the *game design*, during production.

Nevertheless, in the worst case, the *Software Developer* would not be in charge of the *game design*; the financial backers would. And the time it would take to edit the software, after the *game design* had been changed by the financial backers, would not be negligible.

A solution that could address this worst case would also benefit the better cases. It would benefit those *Developers* who did not have the resources to produce and maintain a complete *game design*, during the production process, to meet the demands of the financial backers. And it would also benefit those *Developers* who did have the resources. Since they could save these resources (that is staff, time and money), if they choose to.

So, in the worst case, another way in which this definition for the language of the staff could be provided for would be through a *software architecture*.[1] And through a software production process based on that *architecture*. The *software architecture* would be used to produce *technical designs* that would meet the requirements of the *game design*, as these changed throughout production. The *architecture* would provide principles for how the *software designs* (i.e. *technical designs*) were composed and changed. So that, the *software designs* would facilitate the definition of the language used by the staff to communicate.

The software production process, based on the *architecture*, would be used to ensure that *architecture* was applied correctly, from the beginning to the end of production. It would verify that each *software design* produced met both the

DOI: 10.1201/9781003502784-2

requirements. That is to say the requirements of each change in the *game design* and the requirements of the *software architecture*. Thus, the production process would also ensure that the software itself always kept its definition. That is, it never degenerated, but always kept the same structure and did not increase in complexity.

These two solutions are namely the **EVENT-DATABASE ARCHITECTURE** and the **EVENT-DATABASE PRODUCTION PROCESS**.

In the **Event-Database Architecture**, the two principles that would govern the composition and the changing of a *software design* would be a set of **EVENTS** and a **GAME DATABASE**.

Events would control the flow of the game. Any external or internal *branch*[2] in its behaviour would be caused by an **Event**. Each **Event** would have an identifier (or *ID*[3]). This would be either a number or a word. Each **Event** would have one or more **ACTIONS**, which would respond to it. An **Action** would be one or more *software procedures*,[4] which would be performed in response to an **Event**. An **Action** could either be as small as simply playing a sound, or displaying a single image. Or it could be as large as loading or starting in a new part of the *Game World*.

Actions would be performed by separate *game modules* (or **GAME OBJECTS**). Each **Game Object** would be simple and perform a small set of **Actions**. There would be lots of these, as well as many **Events**. This would give the *Game Producers* and *Game Designers* the option of inserting items into, or modifying, the flow of the game as they liked. This would allow them to adapt the game to their vision, as this changed. But much more than that, **Events** would provide a means of identifying relationships between the different components of an incomplete *game design*. And by doing so, these numbers or words would contribute to the definition of the language, for all the staff building, or extending that *game design*.

The *Database*[5] part of this *software architecture* would be a single, central storage of information, which would hold all the *data* used in the game. This includes the shared *Game data*, from the *technical design*, and the *Abstract data*, from *game modules* and *software libraries*.

This **Game Database** would be a *Relational Database*.[6] A *Relational Database* was originally designed to manage large amounts of business *data*. But it has now become the most widely used form of *Database* for many different types of applications.

A *Relational Database* holds *data* about a set of entities. An entity is a single word or abstract concept in a field of application. In this case, the field of application is Computer Games. So each entity could be

a character in the *Game World*,
a location,
an animate or inanimate item in a location,
the animation of a character or an item,
a menu,
a button on a menu,
a piece of music,
a sound effect,
a graphical effect,
and so on.

A *Database* is made up of *Database Tables*.[7] Each group of entities (characters, locations, animate or inanimate items and so on) is held in the same *Database Table*. And each *Database Table* is made up of rows known as *Database Records*,[8] and columns known as *Database Fields*.[9]

A *Database Record* holds the *data* for a single entity in a *Database Table* e.g.

the name of a character,
its location in the *Game World*,
its health,
its inventory,
and so on.

A *Database Field* holds a single piece of *data* about each entity. For example, a *Database Field* which holds the names of characters could hold

Matthew
Mark
Luke
John
James
Simon

The first *Database Field*, in each *Database Record* is used to identify that *Record* in the *Database*. And it is known as a *Primary Key*.[10]

The *Database*, used in the **Architecture**, would be created and maintained by a professional, *Database Administrator*.[11] The *Administrator* would maintain the quality of the description of the *data* at a high standard. The *Administrator* would also ensure that there were no duplications of data in the *Database*. This, along with the high quality of the description of that *data,* would provide the staff with a high-quality language to use. The quality of this language would be maintained by the *Database Administrator* from the beginning to the end of the production process.

The *Database Tables* would fall into four main categories:

1. publicly shared *Game data*
2. privately restricted *Abstract data* that only specific *software modules* or *software libraries* would use
3. the properties of **Events**
4. the properties of **Game Objects**

You may be tempted to break up the single **Game Database** into four or more smaller *Databases* because

a. there are four categories of *Database Tables*.
b. access to some of the *data* e.g. private restricted *Abstract data* should be hidden and restricted.

But in doing so you may, in turn, tempt some of the staff to vary the quality of description for each of these smaller *Databases*. Once some of the *Databases* were out of sight, it would be very easy for these to become neglected. Or for some of the smaller *Databases* to be treated as somehow special. And therefore not in need of the same quality of description as the rest of the **Game Database**.

However, as already mentioned, the high quality of the description of the *data* in the **Game Database** produces a high-quality language for the staff to use to communicate. And if the quality of the description of one of the smaller *Databases* degenerates, then the communication amongst the staff will degenerate. And you will start noticing *private conversations* occurring only between the staff who understand the degenerative language of that *Database*.

In certain situations, hiding private restricted *Abstract data* may simplify software. This may reduce unnecessary dependency between *software modules*. If one or more *software modules* depend on the internal structure of the private restricted *Abstract data*, of another, it increases the complexity of changing that structure. Whenever you modify that internal structure, then you have to modify all of the *software modules* which access it. Therefore, you may see a need to have a separate *Database* for this private restricted *Abstract data*.

But this increase in complexity only arises if the staff who write a new *software module* write it in such a way that it depends on the internal structure of the private restricted *Abstract data* of another. However, the perception of the description of that private restricted *Abstract data*, by the staff, does not increase the complexity. Nor does placing all *Abstract data* in one central **Game Database**.

By placing all the *data* in one central **Game Database** you would ensure that the quality of the description of the *data* was consistent throughout.

Another important requirement of **Game Database** is that it should be in a non-proprietary, *open data format*.[12] This gives you the following advantages:

1. You do not restrict yourself when it comes to your choice of software you use to create and manage the **Game Database.**
2. You do not restrict yourself if you choose to outsource the creation and management of the **Game Database** to another company.
3. You can scale up the **Game Database** to the point where you can outsource the creation and management of it to another company.
4. If two *platforms* support the **Event-Database Architecture**, you can quickly transfer the game built with the **Architecture**, from one *platform* to another, by transferring the **Game Database.**
5. You can modify the **Game Database** using a wide variety of tools, from in-house custom tools to off-the-shelf third-party products.

This last advantage is the most important. It follows on from the fact that the descriptions of the *data* in the **Game Database**, to a large extent, will define the language of the project. And hence it should be understood by as many people as possible. The *open data format* requirement provides the ultimate test of this understanding. If the quality of the descriptions were high, then different staff using different tools will be able to modify it to effect changes in the *game design*. Using the description of the *data* in the **Game Database**, and any tools which the staff were familiar with which could interoperate with the *Database* through its *open data format*.

It may be tempting to assume that, because the two main principles of the **Event-Database Architecture** were simple or familiar (i.e. a set of **Events** and a **Game Database**) that producing *technical design* based on it would be trivial. But it is not. The natural tendency of building a *technical design*, using very simple principles, is to produce complex results. The *Software Evolution Process* perfectly illustrates that.

It too has two seemingly very simple principles. Let the software grow slowly and evolve over the time. And let the basis of this evolution be feedback from the software user. And yet, whenever you use the *Software Evolution Process* on a project in the Computer Games industry that involves 60 or more staff, it produces very complex results every time. For getting it to produce simple results is not trivial. And that is what the **Event-Database Production Process** will attempt to do, with the **Event-Database Architecture**, as you shall see later on.

NOTES

1. *Software architecture.* A description of a system for producing software. It includes a description of the components of the system, the relationship between these components, and the principles that govern how these components change. See Glossary.
2. *Logic branch.* The point where a software procedure decides to follow one path or another, in its overall task.
3. *ID.* Identifier. A word which identifies one or more set of data. In a Database, the ID of each Record is a special Field known as the Key Field or Primary Key.
4. *Software procedure.* A sequence of instructions, for a computer, to perform a task. The sequence can be used again and again to repeat that task.
5. *Database.* A collection of data arranged for ease and speed of search and retrieval.
6. *Relational Database.* A Database where all the software data, and the relationship between these, are organised in tables with rows, known as Database Records, and columns, known as Database Fields.
7. *Database Table.* A collection of Database Records. A group of related data about entities (e.g. characters, locations, or items in a Game World) which share the same properties in a Relational Database.
8. *Database Record.* A collection of Database Fields. A group of related data about a single entity in a Relational Database (e.g. the name of a character in the Game World, its location, its health, its inventory).
9. *Database Field.* A single property of an entity in Database (e.g. the name of a character in the Game World). An element in a Database Record.
10. *Primary Key.* The first Database Field of a Database Record that is used to identify that Record, search for it and refer to it. This has to be a unique word or number.
11. *Database Administrator.* A company employee who is responsible for the design and management of one or more Databases. The employee is also responsible for the evaluation, selection and implementation of the Database management system.
12. *Open data format.* The description of the layout of data in a Database, and how each data is used. This description is freely available for all software applications to use to read and modify the Database.

3 The Software Architecture

A set of **Events** and a **Game Database** by themselves would not be enough to produce a game. A host of *software modules* would be required to enable these to present a *User Interface*, which the player could use to interact with the game. These *modules* would need to support the addition of a large number of **Game Objects**, which may be added to the **Event-Database Architecture**, to customise it for a particular game. The *modules* would need to send **Events** to the correct **Game Objects**. And it would also be useful, but not necessary, for the *modules* to provide a basic model for moving visible **Objects**, in the world where the game takes place.

In the complete **Event-Database Architecture**, there would be seven of these **HOST MODULES**, that would provide all of these services. And there would also be an eighth one, which would control and synchronise all of these *modules*.

The first two would be, of course, an **EVENTS HOST** and a **DATABASE HOST**. The rest would be an **OBJECTS HOST**, a **PHYSICS HOST**, a **GRAPHICS HOST**, a **SOUNDS HOST**, a **GAME CONTROLLERS HOST** and a **CENTRAL HOST**.

3.1 EVENTS HOST

The **Events Host** would direct all **Events** or chains of **Events** in game. For each **Event** it received, it would direct the **Events** that should follow from it. There would be two types of **Events** the **Host module** would direct: **PRIMARY EVENTS** and **SECONDARY EVENTS**.

Primary Events would be the **Events** that the **Host module** received directly from the **Game Objects** or the other **Host modules**. **Secondary Events** would be the **Events** it sends to the **Game Objects** to perform **Actions**. Each **Primary Event** would cause one or more **Secondary Events** or **Actions**. This would allow the *Game Producers*, *Game Designers* and other staff to create and control the chain of **Events** in a game. This would also allow them to re-use any **Action** of a **Game Object**, in response to an **Event**, in any other **Events** that happened during a game.

For example, if two different **Primary Events** required the same response or **Action** from a **Game Object**, then the *Game Designers* could map both **Primary Events** to the same **Secondary Event** or **Action**.

During the production process, it may be that two different groups of staff may be adding two features to a game using two different **Secondary Events**. And it may turn out that these two features conflict with one another and overlap. That is, both **Secondary Events** follow on from the same **Primary Event**. And instead of removing one feature the staff want to keep both features in that chain of **Events**, but

 DOI: 10.1201/9781003502784-3

mutually exclusive at the same time. That is, only one of the two should follow on from the same **Primary Event** at any given time.

In these cases, each **Secondary Event** will have a priority that the **Events Host** will use to determine which of the two **Secondary Events** should follow on from that **Primary Event**. **Secondary Events** with the same priority will have an equal chance of following on from that **Primary Event**. **Secondary Events** with a higher priority will have a higher chance of following on from that **Primary Event**.

Whatever chain of **Events** were generated during the history of a game, the **Events Host** would add each **Primary Event** it received, and **Secondary Events** it generated, to a history of that chain in an **EVENTS HISTORY RECORD**. Before responding to that **Primary Event** or passing on that **Secondary Event**. This would include the following *Database Fields*:

1. a *Primary Key*
2. the maximum number of different types of **Events**
3. the maximum length of the history of **Events**
4. the history of **Events**

This history would include all the **Primary** and **Secondary Events** in chronological order. The maximum length of the history would be determined by the needs of the *game design* and the number of **Events** that it expects to be executed from the beginning to the end of that game.

This history would have several advantages. These would include

1. allowing players or computers who join multiplayer games late to replay all of the **Events**, since the beginning of the game, to synchronise the local copy of the *Game World* with the rest of the copies on the network
2. allowing **Game Objects** to perform **Actions** in response to **Events** that change depending on antecedent **Events**
3. allowing *Game Testers* to diagnose steps needed to reproduce *Bugs*, critical errors or Crashes
4. allowing *Game Programmers* to diagnose the code executed to reproduce *Bugs*, critical errors or Crashes

So long as these *Bugs*, critical errors or Crashes do not remove the history. This means the **Events History Record** should not be reset. Even if the game was shut down or restarted manually by a player or automatically by an error. It should only be reset when the *Game World* is reset.

Each *technical design*, based on the **Event-Database Architecture**, would strike that balance. It would define the maximum different types of **Events** there would be in the production process. It would define the maximum length of the history of **Events** that should be kept. It would define whether the history would contain temporal or spatial information. And it would define whatever set of **Primary Events**, **Secondary Events** and priorities of **Events** were needed to meet the requirements of the *game design*. But this set would always include the following standard **Primary Events**, which would be required by the **Host modules**, in chronological order:

- *PRIMARY INITIAL RESET EVENT.* This would be sent initially by the **CENTRAL HOST**, a **Host Module** that sets up and synchronises all the other **Modules**, after it has finished setting these up and before the game starts. This **Event** would also be sent subsequently, by any **Game Object**, to restart the game from any point along its flow.

- *PRIMARY CONNECT EVENT.* This would be sent by the **Game Controllers Host**, when a *Game Controller*[1] was connected to the computer hardware. For example, when a player joins a game.

- *PRIMARY CONTROLLER MOVED EVENT.* This would be sent by the **Game Controllers Host**, when a device on a *Game Controller* started to move. For example, when a player issued a command for the player's character to start walking in the *Game World*.

- *PRIMARY COLLISION EVENT.* This would be sent by the **Physics Host** when a **Game Object** was involved in a collision with another **Object**. For example, when the player's character walks into another character in the *Game World*.

- *PRIMARY PROXIMITY EVENT.* This would be sent by the **Physics Host**, when a **Game Object** had come within, or moved beyond, close proximity of another **Object**. For example, when a player's character walks within close proximity or moves beyond another character.

- *PRIMARY END EVENT.* This would be sent by any **Host Module** or **Game Object** whenever the playback of a recorded sequence came to an end. For example, this would be sent by the **Sounds Host**, when a *sound stream*[2] had finished playing a player's character's footsteps. Another example, this would be sent by a **Game Object** when it had finished playing an animation sequence of the player walking. And if that sequence were part of a longer sequence, it would mark the end of one part of the animation and the beginning of the next.

- *PRIMARY CONTROLLER STOPPED EVENT.* This would be sent by the **Game Controllers Host**, when a device stopped moving. For example, when the player issued a command for the player's character to stop walking in the *Game World*.

- *PRIMARY CONTROLLER PRESSED EVENT.* This would be sent by the **Game Controllers Host**, when a button on a *Game Controller* was pressed. For example, when the player issued a command for the player's character to start jumping in the *Game World*.

- *PRIMARY CONTROLLER RELEASED EVENT.* This would be sent by the **Game Controllers Host**, when a button was released. For example, when the player issued a command for the player's character to stop jumping in the *Game World*.

- *PRIMARY DISCONNECT EVENT.* This would be sent by the **Game Controller's Host**, when a *Game Controller* was disconnected. For example, when the player left the game.

- *PRIMARY SHUTDOWN EVENT*. This would be sent only once, by any **Game Object**, when it was time to exit the game. When the **Events Host** received this it would shut down itself down, after executing all the **Secondary Events** associated with that **Primary Event**.
- *PRIMARY PROJECTION EVENT*. This would be sent by the **Graphics Host** after it selected the items that would be rendered in the *Game World*. And before these items were rendered. And any **Game Object** could respond to this **Event** and change the items being rendered on the screen.

And it would include the following set of standard **Secondary Events** required by the **Host Modules**:

- *SECONDARY CONNECT EVENT*. This would be received by a **Game Object** when a device on a *Game Controller* was connected to the computer.
- *SECONDARY CONTROLLER MOVED EVENT*. This would be received by a **Game Objec**t when a device started to move along an axis.
- *SECONDARY CONTROLLER STOPPED EVENT*. This would be received by a **Game Object** when a device stopped moving.
- *SECONDARY CONTROLLER PRESSED EVENT*. This would be received by a **Game Object** when a device was pressed.
- *SECONDARY CONTROLLER RELEASED EVENT*. This would be received by a **Game Object** when a device was released.
- *SECONDARY DISCONNECT EVENT*. This would be received by a **Game Object** when a device was disconnected.

Of all these standard **Events**, the **Primary Initial Reset Event** would be the most significant. Since this would distinguish most clearly any production process, based on the **Architecture**, from a *Software Evolution Process*.

For, as has already been explained, during a *Software Evolution Process* in the Computer Games industry, the vast majority of the *game modules* would not be documented. Hence, the *Game Programmers* would not be aware of how many and which *game modules* were involved at any point in the change of the flow of the game. This includes the beginning and end of any stage or level in the game. This would include the very first stage: the beginning of the game.

The consequence of all this is that it would be impossible to reset the game, back to its beginning, from any point along its flow, without incurring errors.

However, in the **Event-Database Architecture** the **Primary Initial Reset Event** should do exactly that. It should reset the game back to its beginning, from any point in the flow of the game. And you should be able to replay the game from the beginning. If you cannot replay or repeat the game, then that means an error has occurred and the **Primary Initial Reset Event** has failed. If a game based on the **Event-Database Architecture** cannot pass this test, then it is not following the **Architecture**.

The properties of the **Primary Initial Reset Event**, and all other **Primary** and **Secondary Events**, would be held in the *Records*, of the **Game Database**. For each **Primary Event**, this would be a **PRIMARY EVENT RECORD**. This would include the following *Database Fields*:

1. a *Primary Key*
2. the list of all **Secondary Events** which would be sent in response to that **Primary Event**.

Each **Game Database** would have at least two of these *Records*. These would be the **INITIAL RESET EVENT RECORD** and the **SHUTDOWN EVENT RECORD**. The former would hold a list of **Secondary Events** for the **Initial Reset Event**. And the latter would hold those for the **Shutdown Event**.

For each **Secondary Event**, there would be a **SECONDARY EVENT RECORD**. This would include at least the following *Database Fields*:

1. a *Primary Key*
2. the time delay before the **Secondary Event** would be sent after a **Primary Event**
3. the *game time*[3] it was due to be sent
4. the **Game Object** that would receive it
5. the *Primary Key* of another *Record*, containing the list of any other, possibly overlapping **Secondary Events**, over which it had a higher or lower priority
6. the priority of the **Secondary Event**
7. a list of one or more **Game Objects** which caused that **Event**.

The **Events Host** would send **Secondary Events** to **Game Objects** through the **Objects Host**. Each **Event** would either be sent immediately or after a set time. The exact time would be defined in the properties of each **Secondary Event**.

Before the **Events Host** sent an **Event** with a time delay, it would add the delay to the current *game time*. It would insert the result into the *Field* which held the time that the **Event** was due to be sent. It would then add that **Event** to a list. This list would be part of a **DELAYED EVENTS LIST RECORD**. This would have the following *Database Fields*:

1. a *Primary Key*
2. a list of the Keys for **Secondary Events**.

The list would be ordered by the *game time* each **Secondary Event** was due to be sent.

After sending all the **Events**, with no time delays, the **Events Host** would check the *Database Table* holding all the delayed **Events**. If the time for any of these were due, it would be sent and removed from its **Delayed Events List Record**.

Any set of **Secondary Events** which could overlap would be described by a **PRIORITY EVENTS LIST RECORD**. This *Record* would have the following *Database Fields*:

1. a *Primary Key*
2. an ordered list of Keys of **Secondary Events**, ordered by priority
3. an ordered list of priorities for each **Event**.

So that the priority of each **Event** in the first list would be its respective number in the second list.

Any two **Secondary Events** would only be considered to overlap if these occurred in response to the same **Primary Event**. And these two were on the list of the same **Priority Events Record**.

When a **Primary Event** occurred which had two or more overlapping **Secondary Events**, the **Events Host** would add up all the priorities of all the **Secondary Events**. It would then divide this total into intervals, one interval for each **Secondary Event** and the size of each interval would be the size of the priority of that **Event**. So the size of the first interval is the size of the highest priority **Secondary Event**. And the size of the second interval is the size of the second highest priority and so on. It would then choose a random number between 0 and the total size of the priorities. And depending on which interval this number fell, it would select the **Secondary Event** assigned to that interval as the **Event** which would respond to the **Primary Event**. The effect would be that the probability of a **Secondary Event** being selected would be proportional to the relative size of its priority. There is a diagram showing the flow of information to and from the **Events Host** in Figure 3.1 with a Legend in Figure 3.2.

3.2   DATABASE HOST

The **Database Host** would store and retrieve information from the **Game Database**. It would allocate all the space in the computer memory required to store this *Database*. And it would release this space when the game was shut down. This *Database* would include the properties of all **Events**, all shared *Game data*, the properties of all **Game Objects**, any *Abstract data* these **Objects** used and any *Abstract data* the other seven **Host modules** used.

As well as a *Record* for each of these items, the **Game Database** would hold other *Records*, especially for the **Database Host**. These *Records* would be for analysing the access of *data*.

The first of these would be a **DATABASE MONITOR RECORD**. This would describe the set of *data* whose access would be monitored by the **Database Host**. And the second would be a **DATABASE LOG RECORD**. This would contain the details of the access to that *data*.

A **Database Monitor Record** would contain the following *Database Fields*:

1. a *Primary Key*
2. a list of *Primary Keys*, of the *Records* whose access would be monitored
3. a list of *Primary Keys* of **Database Log Records**.

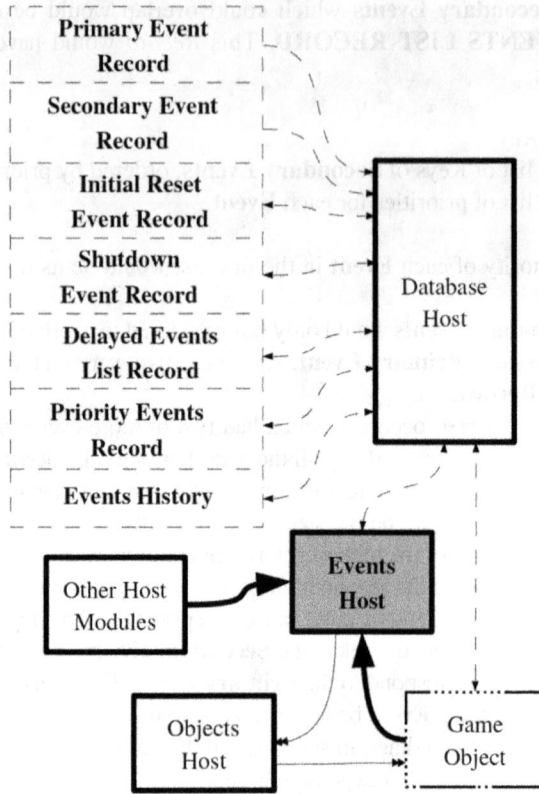

FIGURE 3.1   The flow of information to and from the **Events Host**.

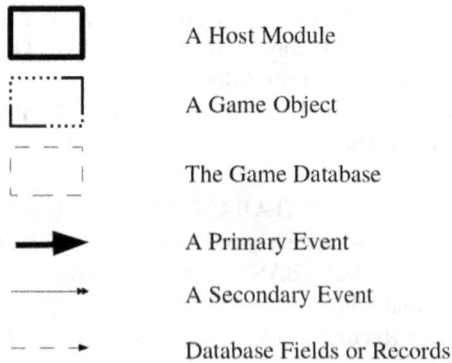

FIGURE 3.2   Legend of the symbols displayed in Figure 3.1. It is a list of the symbols, for the components of the **Event-Database Architecture**, that would interact with the **Events Host**.

Each monitored *Record* in the first list would have a corresponding **Database Log Record** in the second list. And every change in a monitored *Record* would be kept in its **Database Log Record**.

Each **Database Log Record** would have the following *Database Fields*:

1. a *Primary Key*
2. a list of times the monitored *Record* was modified
3. a list of the names of the modified *Fields*
4. a list of the old values in each *Field*.

Each modification time in the first list would have a corresponding entry in the second list. And each name of a modified *Field* in the second list would have a corresponding entry in the third list. The third list would contain the original values of the modified *Fields* before these were modified.

When the game was shut down, the **Database Host** would write all **Database Log Records** to a file, next to the **Game Database**. This could then be analysed, for example, to find out when an error occurred in the *data* when playing a game.

But apart from debugging purposes, the **Database Log Records** would have a direct bearing on delayed **Secondary Events**. Any **Game Object**, responding to a **Secondary Event**, could use the **Database Log Records** to look back in time, at the set of *data* which caused any of the preceding **Primary Events.** Even if that *data* had changed by the time the **Game Object** responded.

This could be done if the *Database Records* the **Game Object** wanted to look back in time at were being tracked by the **Database Monitor Record**. And each tracked *Record* had a **Database Log Record** assigned to it. And the **Game Object** had a *Database Field* with the list of the Primary Keys of these **Database Log Records**. So that it could pass these Keys to the **Database Host**, to get the **Database Log Records** and search through the timeline in the **Database Log Records** for the value of the *Fields* in the tracked *Records*, at some earlier point in time.

As well as tracking modifications to the values of *Database Fields*, the **Database Host** would also track the computer memory it was using. The **Database Host** would have a **RESIDENTS LIST RECORD.** This would have the following *Database Fields*:

1. a *Primary Key*
2. an ordered list of all of the *Primary Keys* of *Records* which were currently loaded into the computer memory, in the order these were loaded
3. a maximum length of this list

This would be used to keep track of the *Database Records* it had loaded into computer memory, excluding those which were either not loaded or were stored in the storage media holding the **Game Database**. Once this number exceeded the maximum, the **Database Host** would unload some *Records* to make space for more, using some suitable steps or algorithms. For example, unloading the oldest *Records* residing in memory or the least frequently used *Records*.

The **Database Host** would also have an **ABSENTS LIST RECORD.** This would have the following *Database Fields*:

1. a *Primary Key*
2. an ordered list of all of the *Primary Keys* of *Records* which were unloaded from the computer memory, in the order these were unloaded
3. a maximum length of this list

This would be used to keep track of the *Records* which were loaded into computer memory and were subsequently unloaded for some reason. For example, to save space in computer memory or to write it to some temporary location in a local or remote storage media. From which it could be fetched at a later point in time.

In summary, the **Database Host** would have at least seven *software procedures* which would be used to read, write and query *Records*. These would do the following:

1. get a single *Record* from the **Game Database**, when given its Primary Key, adding it to the **Residents List Record**, and if the list exceeds its maximum length, unloading the oldest *Record* in the list from computer memory and adding it to the **Absents List Record**
2. get multiple *Records*, when given a list of *Primary Keys*
3. get multiple *Records* whose *Fields* holds values that exactly match a given value
4. get multiple *Records* whose *Fields* holds values that match one of a set of given values
5. get multiple *Records* whose *Fields* holds values that fall between a maximum and minimum given value
6. modify a *Record* in the *Database*, when given its Primary Key and a new *Record*
7. get the previous version of a modified *Record*, when given its Primary Key and the *game time* before it was modified. But only if that *Record* were being tracked by the **Database Monitor Record** and had a **Database Log Record**. Otherwise, this *procedure* should just get the latest version of a modified *Record*

There is a diagram showing the flow of information to and from the **Database Host** in Figure 3.3 with a Legend in Figure 3.4.

3.3   OBJECTS HOST

The **Objects Host** would set up and direct all the *game modules* (or **Game Objects**). It would allocate all the space in the computer memory required to load and set up the **Game Objects**. And it would release this space when the game was shut down. After allocating the space, the **Objects Host** would load and set each **Object** up with its *Record* in turn. The number and order in which these would be loaded and set up would be determined by the **Game Database**.

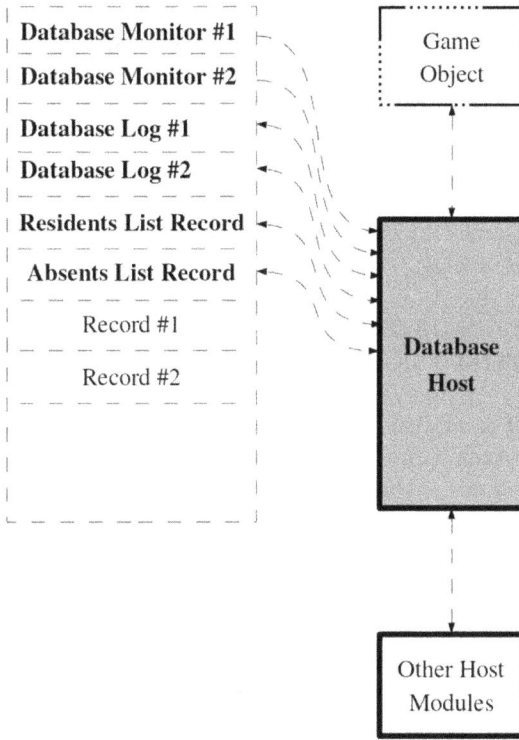

FIGURE 3.3   The flow of information to and from the **Database Host**.

This would be the **OBJECTS LIST RECORD** with the following *Database Fields*:

1. a *Primary Key*
2. an ordered list of *Primary Keys*, of **GAME OBJECT RECORDS**.

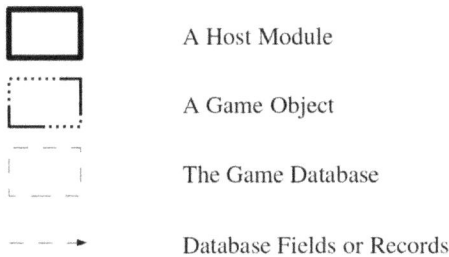

FIGURE 3.4   Legend of the symbols displayed in Figure 3.3. It is a list of the symbols, for the components of the **Event-Database Architecture**, that would interact with the **Database Host**.

Each **Game Object** would have one **Game Object Record**. The **Game Object Record** would have the properties of the **Game Object** or **GAME OBJECT ATTRIBUTES** in its *Database Fields*. The number and type of *Fields* would vary depending on the **Game Object Attributes.** These include physical attributes such as the location, size, boundaries and appearance of the **Game Object** if visible in the *Game World*. And these include its non-physical attributes such as the **Events** or sounds it generates.

Whatever the **Game Object Attributes** were each *Record* would have at least the following *Database Fields*:

1. a *Primary Key* or **OBJECT ID**
2. a **GAME OBJECT CODE FIELD**

The **Object ID** is an identifier of the **Game Object**. It is used throughout the **Event-Database Architecture** to refer to that **Object**, either by other **Game Objects**, **Host Modules** or **Database Records.**

The **Game Object Code Field** would control how **Game Objects** would respond to **Events**. It is not right for a high level tool like a *software architecture*, such as the **Event-Database Architecture**, to concern itself with low level tools, like the machine code instructions for **Game Objects**. And the **Game Object Code Field** is not concerned with these low level tools. It is an abstraction of a **Game Object** and it can be as high level or as low level as you want.

The **Game Object Code Field** would either hold a unique code word or number or name of a file which identifies to the **Objects Host** which **Game Object** to load into computer memory. And then pass the Primary Key of the *Record* to that **Game Object**, for it to use to set itself up. Or the *Field* would hold the actual instructions, either machine code or virtual machine code, for the **Game Object**. This can be in the form of a *software library* that the **Objects Host** will load into the computer memory. But whereas a normal *software library* may contain several *software modules*, this *software library* will only contain the code for one and only one *software module* i.e. the **Game Object**.

The advantage of the **Game Object Code Field** only holding a code word or number that identifies the **Game Object** to be loaded is that the *Field* will not have to store a large amount of *data*. And the contents of the *Field* is independent of the computer hardware. The disadvantage is that you need another external data structure that maps the code number to the instructions of the **Game Object** which is outside of the **Game Database**. And therefore the description of that data structure may not be the same quality as the description of the **Game Database**.

The advantage of the **Game Object Code Field** holding the actual instructions, either machine code or virtual machine code, for the **Game Object** is that you do not need an external data structure. The **Objects Host** can simply load the instructions for the **Game Object**, from its **Game Object Record**, to computer memory. And then pass the *Primary Key* for that *Record* to the **Game Object**. The disadvantage is that the *Field* may be very large depending on the number of instructions. And the contents of the *Field* is dependent on computer hardware if it is machine code, instead of virtual machine code.

The advantage of the **Game Object Code Field** holding the name of a file which identifies the **Game Object** is that *Field* again does not have to store that large amount of *data*. Instead the file could contain the machine code or virtual machine code for the **Game Object** that would be loaded into memory. Or the file could contain the original programming language instructions. That the **Objects Host** would translate or 'compile' into machine code before loading it into memory. And in this way the **Objects Host** would dynamically 'compile' and load the **Game Objects** into memory as and when needed. So the game could be restarted very quickly after changes were made to the programming language instructions of the **Game Objects**. And the game need not be restarted at all if the **Game Object** has not been loaded yet. Or if the **Objects Host** detects when the file has been modified and automatically unloads it from memory, rebuilds or re-'compiles' the **Object** and reloads it into memory.

The disadvantage is that it would take longer for the **Objects Host** to 'compile' and load the **Game Objects** into memory the first time. And the programming language instructions have to be written to 'compile' on as many different computer hardware that the game will run on. And some computer hardware may not actually allow you to host a 'compiler tool' that could 'compile' the code.

After each **Game Object** had been loaded and set up with its **Game Object Record**, the **Objects Host** would then start to direct any appropriate **Events**, from the **Events Host**, to that **Object**.

The **Objects Host** would have one *software procedure* which would be used to receive **Events**. The **Events Host** would use this *procedure* to provide the **Objects Host** with each **Secondary Event**, and its **Game Object** (i.e. The Primary Key of that **Secondary Event** and the Primary Key of that **Game Object**).

Each **Game Object** would, likewise, have one *procedure* which would be used to receive **Events**. The **Objects Host** would use it to provide that **Object** with each **Secondary Event**. When it received an **Event**, each **Object** would use one of its other *procedures* to execute each **Action** in response, depending on its role in the overall *game design*.

The **Objects Host** would have two other *Records* in the **Game Database**. One would be the **OBJECTS FAILED LIST RECORD.** This would contain the following *Database Fields*:

1. a *Primary Key*
2. a list of *Primary Keys* of **Game Objects** that were used to execute **Actions** but failed.

Another would be **OBJECTS FAILED TIMES LIST RECORD.** This would have the following *Database Fields*:

1. a *Primary Key*
2. a list of times that each **Object** failed when executing an **Action**.

Before executing an **Action** on a **Game Object** in response to an **Event**, the **Objects Host** would add the Primary Key for that **Object** to the **Objects Failed**

List Record. And it would add the current time to the **Objects Failed Times List Record**.

If the **Action** were executed successfully, then the Primary Key of that **Object** would be removed from the list. And the corresponding time would be removed from the **Objects Failed Times List Record**.

If the **Action** failed to execute, and the **Objects Host** had a critical error or Crashed executing that **Action**, then the **Object** and the time would remain on those two lists. Even when the game was restarted.

When the game was restarted after a Crash, if the **Object Host** received a **Secondary Event**, for a **Game Object** that was on the **Objects Failed List Record**, before the game was restarted, it will not execute any **Actions** on that **Object**. The **Objects Host** will continue to ignore that **Game Object**, until the error that occurred had been fixed by the *Game Programmers*. And they have manually edited the **Objects Failed List Record** and removed the Primary Key of that **Object** from the list and the corresponding time from the **Object Failed Times List Record**. This means the **Objects Host** and the game will automatically recover from small errors in individual **Game Objects**. Allowing the staff to continue to play and test the rest of the game despite the presence of errors. And at the same time allowing these errors to be highlighted.

To highlight failed **Game Objects**, to the staff, the **Graphics Host** will render any 2D or 3D **Game Object** on that **Objects Failed List Record** in a special over-laying colour. The **Game Object** would be rendered with a red overlaying colour, if the game was started, after the failed **Object** was added to the **Objects Failed List Record**. The **Object** would be rendered in yellow overlaying colour, if the game was started before the **Object** was added to the **Objects Failed List Record**.

The red overlaying colour means the **Object** has a critical error and been dis-abled. That is to say the **Object** has failed and although visible it will be ignored until it is fixed. The yellow overlaying colour means caution. That is to say, the **Object** has caused a critical failure for another player playing the game. But since the current player started playing, before that **Object** failed, the current player may not experi-ence that failure. Unless either they interact with that **Object**. Or they rebuild or get the latest **Game Objects** and restart the game again.

The caution is necessary because the **Game Database** could be used by multiple players in the same *Game World*, across a computer network. And other players, who have not yet encountered a critical error with some **Object**, need to be warned that they may get this error if they interact with it. Or if they rebuild or get the latest **Game Objects** and restart game. However if they do not interact with that **Object**, nor rebuild or get the latest **Game Objects** or restart the game, then they can con-tinue playing.

There is a diagram showing the flow of information to and from the **Objects Host** in Figure 3.5 with a Legend in Figure 3.6.

3.4   GRAPHICS HOST

The **Graphics Host** would display all 2D or 3D items on the computer screen. These would include all texts, 2D images or 3D models.

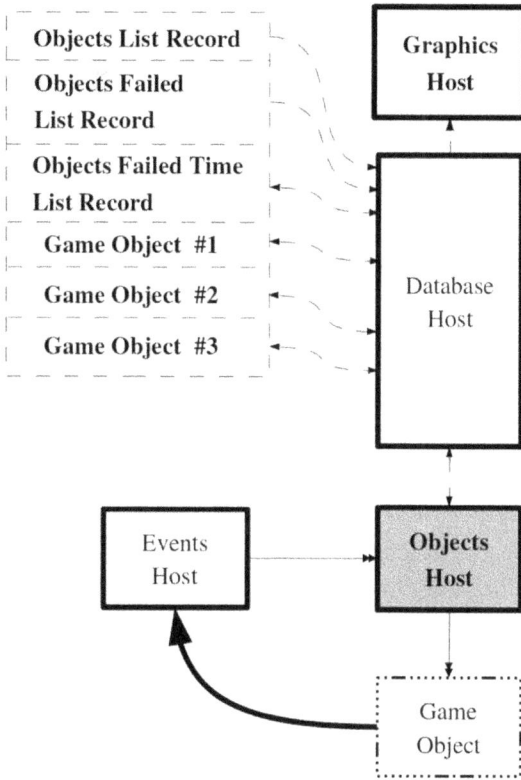

FIGURE 3.5 The flow of information to and from the **Objects Host**.

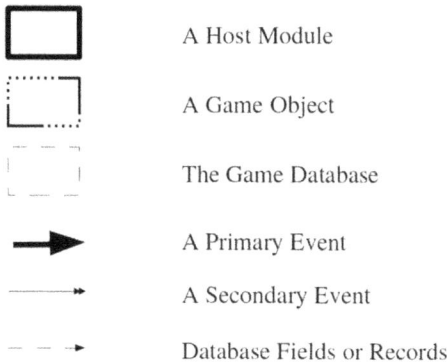

FIGURE 3.6 Legend of the symbols displayed in Figure 3.5. It is a list of the symbols, for the components of the **Event-Database Architecture**, that would interact with the **Objects Host**.

It would require two types of **Game Objects**: a **GRAPHICS OBJECT** and a **CAMERA OBJECT**.

A **Graphics Object** would be a **Game Object** of an item in the *Game World* that could be rendered onto the screen.

A **Camera Object** would be a **Game Object** of a camera in the *Game World*, used to direct a player's view of that *World*, in a 2D or 3D space. This view would be determined by the position, orientation and area of visibility in front of that **Object** or around its position.

The **Graphics Host** would get all its information from *Records* in the **Game Database**. These include

1. the list of **Graphic Objects** whose 2D images or 3D models would be displayed,
2. the shape of each **Graphic Object**,
3. the optional visible surface of each **Graphic Object**,
4. the optional colouring of each **Graphic Object**, and
5. the position and orientation of each **Graphic Object**.

The shape of each **Object** would be described by the list of *vertices*[4] for the *polygons*[5] that make up each 2D image or 3D model.

The surface of each **Object** would be described by the list of *Normal Vectors*[6] of a 2D *polygon* or a 3D model. This may be used for, amongst other things, controlling the lighting of that *polygon* or model.

The colouring of each **Object** would be described by the *Textures*[7] (or images) that would be used to fill its *polygons*. This would also require the *Texture coordinates*[8] for these *polygons*.

However these were described, the shape of each **Object**, its surface, its colouring, its position and orientation would all come from one *Database Record*. This would be the **GRAPHICS OBJECT RECORD**. The number and types of *Fields* this would have would vary from game to game. But it would always have at least the following *Database Fields*:

1. a *Primary Key* or **GRAPHIC OBJECT ID**
2. a **Game Object Code Field**
3. *X Position*[9]
4. *Y Position*[10]
5. *Z Position*[11]
6. *X Angular Position*[12]
7. *Y Angular Position*[13]
8. *Z Angular Position*[14]
9. low resolution **2D POLYGON ID** or **3D MODEL ID**
10. high resolution **2D Polygon ID** or **3D Model ID**
11. **TEXTURE ID**
12. **TEXTURE COORDINATE ID**

The **Graphic Object ID** would be the *Primary Key* of the **Graphic Object Record**.

The **2D Polygon ID** would be the *Primary Key* of a **2D POLYGON RECORD**. The low resolution **2D Polygon ID** would be the **ID** of a low resolution *polygon* used to define the bounding area or bounding box around an **Object**. The high resolution **2D Polygon ID** would be the **ID** of a high resolution *polygon* used to render that **Object**.

The **3D Model ID** would the *Primary Key* of a **3D MODEL RECORD**. The low resolution **3D Model ID** would be the **ID** of a low resolution model used to define a bounding area or bounding box around an **Object**. The high resolution **3D Model ID** would be the **ID** of a high resolution model used to render that **Object**.

The **Texture ID** would be the *Primary Key* of a **TEXTURE RECORD**. The **Texture coordinate ID** would be the *Primary Key* of a **TEXTURE COORDINATE RECORD**.

The **2D Polygon Record** would have the following *Database Fields*:

1. a *Primary Key* or **2D Polygon ID**
2. a list of *vertices* of a *polygon*
3. a list of *Normal Vectors*

The **3D Model Record** would have the following *Database Fields*:

1. a *Primary Key* or **3D Model ID**
2. a list of triangular *vertices* of a 3D model
3. a list of *Normal Vectors*

The **Texture Record** would have the following *Database Fields*:

1. a *Primary Key* or **Texture ID**
2. a width of the *Texture*
3. a height of the *Texture*
4. a list of the colours of the pixels of the *Texture*

The **Texture coordinate Record** would have the following *Database Fields*:

1. a *Primary Key* or **Texture coordinate ID**
2. a list of triangular *Texture coordinates*

To display each *Frame*[15] of the game, the **Graphics Host** will use a combination of *Software Rendering*[16] and *Hardware Rendering*.[17]

It would use *Software Rendering* to perform a quick preliminary projection of **Graphics Objects** on to the screen. After that it would use some criteria to select which of the projections will be rendered fully using *Hardware Rendering*.

The **Graphics Objects** that would be rendered would come from a **GRAPHICS LIST RECORD**. This would have the following *Database Fields*:

1. a *Primary Key* or **LIST ID**
2. a list of **Graphic Object IDs** of **Objects** that should be displayed

The *Software Rendering* will use the Central Processor to project the bounding box or 2D *polygon* or 3D model around each **Graphic Object** in front of the **Camera Object**, from 2D or 3D *Game World*, to the 2D space of the screen. All bounding boxes which were completely outside of the area of visibility or viewing frustum in front, of the **Camera Object**, would be culled.

The results of these projections would be put in the *Database Table* of **PROJECTED SHAPES** or **PROJECTED SHAPES RECORDS**. This would have the following *Fields*:

1. a *Primary Key* or **PROJECTION ID**
2. a list of projected *vertices*
3. a **Graphic Object ID**

Each vertex in the list of projected *vertices* would have three coordinates:

X Position
Y Position
DEPTH COORDINATE

The *X Position* and *Y Position* would determine its 2D position on a *Texture* or the screen. The **Depth coordinate** would determine its relative depth or distance from the **Camera Object**. And whether it was displayed in front of, or behind, other *vertices* with larger or smaller **Depth coordinates**.

After these preliminary projections, some criteria will be used to select which **Objects** would go forwards to *Hardware Rendering*. The *Primary Keys* of **Projected Shapes** selected to go forward would be put in a *Database Table* of **PROJECTED LIST** or **PROJECTED LIST RECORDS**. This would have the following *Fields*:

1. a *Primary Key* or **List ID**
2. a list of **Projection IDs**

The list of **Projection IDs** would be ordered by the distance from the **Camera Object**, using the **Depth coordinates**. So the furthest projection or **Graphic Object** would be at the beginning of the list, and the closest would be at the end.

By default, this would be the order in which the **Graphic Objects** would be rendered on a *Texture* or the screen by the *Hardware Rendering* process. And the *Hardware Rendering* process must not change that order.

But, after placing the items on the **Projected List**, the **Graphics Host** would send a **Primary Projection Event**. So that any **Game Object** that wanted to modify the order or the content of the **Projected List** could do so. By attaching its **Secondary Event** to that **Primary Event**. And modifying the order or the content of the list, depending on the requirements of the *game design*, when it received that **Secondary Event**.

After that, the **Graphics Host** would use the **Projection IDs** in the **Projected List**, to get the **Graphic Object IDs** from **Projected Shapes**. And then use the **Graphic Object IDs** to get

1. **Texture ID**
2. **Texture coordinate ID**
3. high resolution **2D Polygon ID** or **3D Model ID**
4. *X Position*
5. *Y Position*
6. *Z Position*
7. *X Angular Position*
8. *Y Angular Position*
9. *Z Angular Position*

of each **Object** from its **Graphics Object Record**. And it would project each of these through the **Camera Objects**, on to a *Texture* or the screen using *Hardware Rendering* and the Graphics Processor.

The properties of each **Camera Object**, in 2D space, would be held in a **2D CAMERA OBJECT RECORD**. This would have the following *Database Fields*:

1. a *Primary Key* or **2D CAMERA OBJECT ID**
2. a **Game Object Code Field**
3. *X Position*
4. *Y Position*
5. *Z Angular Position*
6. visibility width around position
7. visibility height around position
8. **PROJECTION TARGET FIELD**
9. *X Position* of the centre of projection on screen
10. *Y Position* of the centre of projection on screen
11. projection width around centre
12. projection height around centre

The **2D Camera Object ID** would be the *Primary Key* of the **2D Camera Object Record**.

The **Projection Target Field** would control whether the projection through the **Camera Object** would be made either onto the screen or a *Texture* or nothing at all.

The properties of each **Camera Object**, in 3D space, would be held in a **3D CAMERA OBJECT RECORD**. This would have the following *Database Fields*:

1. a *Primary Key* or **3D CAMERA OBJECT ID**
2. a **Game Object Code Field**
3. *X Position*
4. *Y Position*
5. *Z Position*
6. *X Angular Position*

 7. *Y Angular Position*
 8. *Z Angular Position*
 9. *Near focal length*[18]
10. *Far focal length*
11. *Field of View*[19]
12. **Projection Target Field**
13. *X Position* of the centre of projection on screen
14. *Y Position* of the centre of projection on screen
15. projection width around centre
16. projection height around centre

The **3D Camera Object ID** would be the *Primary Key* of the **3D Camera Object Record**.

There may be two or more **Camera Objects**. And some may be active while others are dormant. The list of the active ones would be held in a **CAMERA LIST RECORD**. In 2D space this would be part of a *Database Table* called **2D CAMERA LIST**. This would have the following *Fields*:

1. a *Primary Key* or **List ID**
2. list of **2D Camera Object IDs**

In 3D space this *Table* would be called **3D CAMERA LIST**. And this would have the following *Fields*:

1. a *Primary Key* or **List ID**
2. list of **3D Camera Object IDs**

Now the criteria for selecting which **Objects** in **Projected Shapes** that would go forwards to the **Projected List** would typically be whether the bounding box of **Objects** were obscured by the bounding box of other **Objects** in front of the camera. But the criteria could be whether the projection of the bounding box of an **Object** were too small to be visible on the screen. Or whether the bounding box of an **Object** were partially visible on the screen. Or whether the bounding box of an **Object** was completely visible on the screen. And thus cull the number of **Objects** rendered using *Hardware Rendering*.

Besides helping *Hardware Rendering*, **Projected Shapes** along with **Projected List** would also be useful for querying the order of **Objects**, in order to build game-play features which involve reflection, refraction or absorption of light.

For example, you could query what **Objects** were behind other **Objects** in the line of sight of the player's character? And refract or alter the position of the **Objects** behind the translucent **Objects** in front to show the effect of refraction.

Another example, you could query what **Objects** were being reflected in a mirror? The surface or *Texture* of each mirror could be created by a projection through a **Camera Object** behind that mirror. So you could query whether the player was reflecting an **Object** across a chain of mirrors in the *Game World*. Suppose the player had to reflect the sun across a chain of mirrors from one place to another in the *Game World*. If the first mirror were placed correctly and were reflecting the sun, then the sun would be placed in the **Projection List** of that mirror. And if the second mirror were placed correctly to reflect the first mirror, then that first mirror would be in the **Projection List** of the second mirror. And so on and so on, for up to, say seven mirrors. Therefore, to see whether the sun was being reflected from the first to the seventh mirror, you would just simply look back through the chain of references in the **Projection List** of each mirror, beginning with the seventh and ending with the first.

Another example, you could query what **Objects** were in the line of sight or vision of Non-Player Characters or NPCs? These would simply be the **Objects** in the **Projection List** of the **Camera Object** in the head of each NPC. Assuming that **Camera Object** was always looking in the same direction that the NPC was facing. And depending on the **Objects** on the List the NPC would attack, talk or otherwise interact with those **Objects**.

Likewise, you could use the **Projection List** to absorb light in the NPC's vision. So, for example, you could have a player who had a special ability like a black hole. The effect of which would be that the light of all **Objects** behind that player were absorbed by the player. And any NPC looking towards that player could not see anything behind that player. To do this, you would simply remove all **Objects** that were behind the player's **Object** in the **Projection List** of the NPC's **Camera Object**. And because the **Objects** were removed from the **Projection List**, the NPC would stop attacking, talking or otherwise interacting with those **Objects**.

Another example, you could query what **Objects** were behind the Mouse Cursor on the screen space? Since **Projected Shapes** would contain the position which each **2D Object** or **3D Object** would occupy in the screen space, you could compare the relative positions of **Objects**. You could compare the position of the Mouse Cursor on the screen with the position of all other **Objects** behind it in the **Projection List**. To find out which **Objects** were underneath the Mouse. And therefore which were being selected by the Mouse and should be highlighted in a colour like yellow or white.

In summary, the order in which **Graphics Objects** would be rendered through the **Camera Objects** would be determined, first, by the order in which the **Graphic Objects** appeared in the **2D Graphics List** or **3D Graphics List**.

Second, by the order in which the **Camera Objects** appeared in the **2D Camera List** or **3D Camera List**.

Third, by the order in which the projections of the **Graphic Objects** appeared in the **Projected List.**

The order in the **Projected List** may not be the same as the order in the **Graphics List**. Since the items in the **Projected List** is a subset of the **Graphics List**. And the order in the latter is arbitrary, but the order in the former is dependent on distance from the **Camera Object**. The **Graphic Objects** on the **Projected List** would be sorted by distance from the **Camera Object**. The furthest away would be at the beginning and the closest would be at the end.

Furthermore, any **Game Object** can customise the content of the **Projected List** when it responds to the **Primary Projection Event**.

Each of these *Database Tables* may hold one list or *Record*. But these may also hold multiple lists or *Records*. The **2D Graphics List** or **3D Graphics List** may hold multiple lists, one for each player. The **2D Camera List** or **3D Camera List** may also hold multiple lists, one for each player. And the **Projected List** may also hold multiple lists, one for each **Camera Object**.

The *Hardware Rendering* process must not change the order in these *Tables*. All the items visible through the first **3D Camera Object** in the **3D Camera List** should be rendered first. And these items should be rendered in the order in which these were listed in the **Projected List**. After that, all the items visible through the second **3D Camera Object** should be rendered second. And these items should be rendered in the order in which these were listed in the **Projected List**. And so on and so on. Until the *Hardware Rendering* process reaches the end of the **3D Camera List**. After that, it should repeat the same process for the **2D Camera List**.

There is a diagram showing the flow of information to and from the **Graphics Host** in Figure 3.7 with a Legend in Figure 3.8.

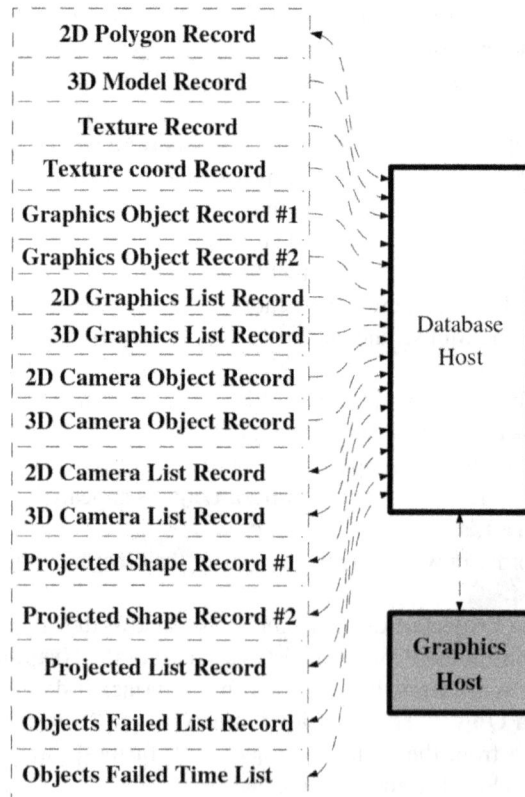

FIGURE 3.7   The flow of information to and from the **Graphics Host**.

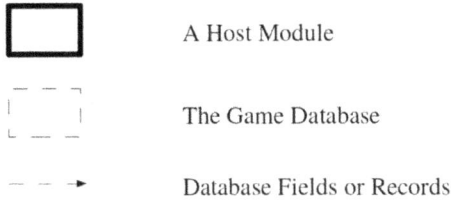

A Host Module

The Game Database

Database Fields or Records

FIGURE 3.8 Legend of the symbols displayed in Figure 3.7. It is a list of the symbols, for the components of the **Event-Database Architecture**, that would interact with the **Graphics Host**.

3.5 PHYSICS HOST

The **Physics Host** would provide a basic model for moving all **Game Objects**, in a 2D or 3D world. It would also detect collisions between the **Game Objects** and when the **Objects** came within close proximity of one another.

The position, speed and acceleration, of each **Game Object**, would be updated using simple Newtonian Physics. The acceleration would be used to update the speed, and the speed to update the position, during each *Unit of game time*.

Collision between **Game Objects** would be detected when the volume of the bounding shapes, around these **Objects**, overlapped. The bounding 2D or 3D shapes can be defined by any 2D *polygon* or 3D model respectively, held in a **Graphics Object Record**.

Collision would also be detected when the line of flight, of a **Game Object**, intersected with a line or a *polygon* of the bounding shape, around another **Object**. The line of flight would be measured from the current position, of the **Object**, to its next position, given its current speed and direction.

Each **Game Object** that would be moved by the **Physics Host** would have two other properties. These would namely be a **Secondary Collision Event** and a **Secondary Proximity Event**. The former would be received by the **Game Object**, when it was involved in a collision. The latter would be received by the **Object** when it was approached (or left behind) by another.

The **Physics Host** would use a **PRIMARY COLLISION EVENT RECORD**. This would have the following *Database Fields*:

1. a *Primary Key*
2. a list of **Secondary Events** of the **Primary Collision Event**

When any number of collisions had occurred, during a *Unit of game time*, the **Physics Host** would prepare. It would begin by clearing all the **Secondary Collision Events**, from the list of **Secondary Events** of the **Primary Collision Event**. For each pair of **Game Objects** involved in a collision, it would add the **Secondary Collision Events** of both, onto this list. Each **Game Object** would then be added to the list of **Objects** which caused the **Secondary Collision Event** of the other. Each one would look up its list to find out which **Objects** it collided with.

After the *Records* for the **Events** had been updated, each colliding **Object** would be placed at the point of impact. The acceleration and speed, of each **Object**, would be modified appropriately to reflect the change of momentum. Finally, a **Collision Event** would then be sent to the **Events Host**, once all the positions of the **Objects** had been updated.

Proximity Events would be handled in the same manner as **Collision Events**. The **Physics Host** would use a **PRIMARY PROXIMITY EVENT RECORD**. This would have the following *Database Fields*:

1. a *Primary Key*
2. the list of **Secondary Events** of the **Primary Proximity Event**.

When any number of **Game Objects** came within, or moved beyond, set areas around other **Objects**, the **Physics Host** would prepare. It would begin by clearing the list of **Secondary Events** of the **Primary Proximity Event**. For each **Object** moving within close proximity of a second **Object**, the **Physics Host** would add the **Secondary Proximity Event** of the second **Object**, onto this list. The moving **Object** would be added to the list of those which caused this **Secondary Event**. The second **Object** would look up this list, from the *Record* of that **Event**, to find out what the moving **Object** was. After the positions of all the **Game Objects** had been updated, a **Primary Proximity Event** would then be sent to the **Events Host**.

The **Physics Host** would get the list of the **Game Objects**, which should be moved, from the **PHYSICS LIST RECORD**. This would have the following *Database Fields*:

1. a *Primary Key*
2. a list of *Primary Keys* of either only 2D **Game Objects** or only 3D **Objects**

The **Host Module** would also get the properties of each **Game Object**, from a **PHYSICS OBJECT RECORD** and update these properties. This would have at least the following *Database Fields*:

1. a *Primary Key*
2. a **Game Object Code Field**
3. a mass
4. *X Position*
5. *Y Position*
6. *Z Position*
7. *X Speed*[20]
8. *Y Speed*[21]
9. *Z Speed*[22]
10. *X Acceleration*[23]
11. *Y Acceleration*[24]
12. *Z Acceleration*[25]
13. *X Angular Position*
14. *Y Angular Position*

15. *Z Angular Position*
16. *X Angular Speed*[26]
17. *Y Angular Speed*[27]
18. *Z Angular Speed*[28]
19. *X Angular Acceleration*[29]
20. *Y Angular Acceleration*[30]
21. *Z Angular Acceleration*[31]
22. low resolution **2D Polygon ID** or **3D Model ID** of a **Graphics Object Record** of the boundary around the **Object** used to test when a **Collision Event** had occurred
23. low resolution **2D Polygon ID** or **3D Model ID** of a **Graphics Object Record** of the boundary around the **Object** used to test when a **Proximity Event** had occurred
24. **Secondary Collision Event** which the **Game Object** should receive
25. **Secondary Proximity Events** which the **Game Object** should receive

The movement of each **Game Object** would be controlled by a force of gravity pulling it down onto a solid surface and a force of friction or viscosity resisting its movement across a solid surface or through a liquid or through a gas or air. This would be applied to its acceleration during each *Unit of game time*, while it had an acceleration and it remained on the **Physics List Record**. The higher the resisting force or acceleration the faster the **Game Object** will slow down and come to rest, after a force or acceleration had been applied to it. And if the resisting force was zero, then the **Game Object** will never slow down.

The force of this gravity and resistance would be enacted by the **MASTER PHYSICS OBJECT**. And its properties would be held in a **MASTER PHYSICS OBJECT RECORD**. This would have at least the following *Database Fields*:

1. a *Primary Key*
2. a **Game Object Code Field**
3. a force of gravity
4. a resisting force or acceleration across solids
5. a resisting force or acceleration through liquids
6. a resisting force or acceleration through gas

There is a diagram showing the flow of information to and from the **Physics Host** in Figure 3.9 with a Legend in Figure 3.10.

3.6 SOUNDS HOST

The **Sounds Host** would play all the sound and music heard during a game. The list or queue of sounds, that were waiting to be played, would come from the **SOUNDS WAITING LIST RECORD** with the following *Database Fields*:

1. a *Primary Key*
2. an ordered list of *Primary Keys* of **SOUND STREAM RECORDS**

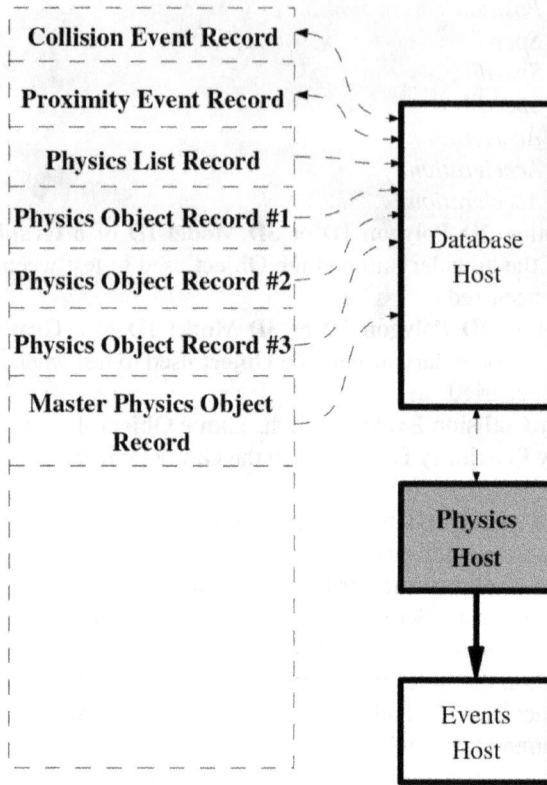

FIGURE 3.9 The flow of information to and from the **Physics Host**.

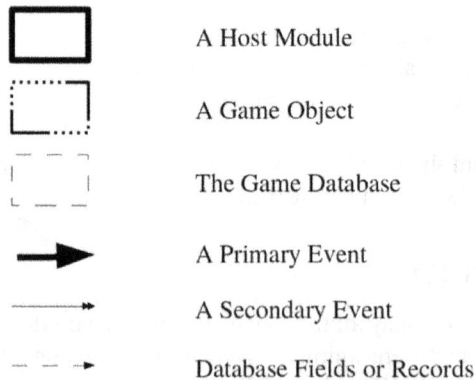

FIGURE 3.10 Legend of the symbols displayed in Figure 3.9. It is a list of the symbols, for the components of the **Event-Database Architecture**, that would interact with the **Physics Host**.

The list or queue of sounds which were being played would also come from the
SOUNDS PLAYED LIST RECORD with the following *Database Fields*:

1. a *Primary Key*
2. an ordered list of *Primary Keys* of **Sound Stream Records**

Each **Sound Stream Record** would have the following *Database Fields*:

1. a *Primary Key* or **SOUND STREAM ID**
2. the *sound stream*
3. its duration
4. the frequency it should be played back
5. *sound channel*[32]
6. left stereo volume
7. right stereo volume
8. **SECONDARY END EVENT**
9. **Object ID**
10. **SOUND RADIUS**

The **Sound Stream ID** would be the *Primary Key* of a **Sound Stream
Record**.

Each *sound stream* would require a **SOUND SPEAKER OBJECT** and a
SOUND MICROPHONE OBJECT to be heard. A **Sound Speaker Object**
would be any **Game Object** that could play or produce a sound. A **Sound
Microphone Object** would be any **Game Object** that can hear a sound, this
includes players.

Each *sound stream* would be either a short sound effect or a long piece of
music that would be heard during the game. Its **Object ID** would be the *Primary
Key* of the **Sound Speaker Object** whose position specified the locality of the
sound or music. Only **Sound Microphone Objects** in close proximity of this
Object would hear the sound. If this **Game Object** were not specified for a
sound stream, then all **Sound Microphone Objects** will hear the sound or
music.

The **Sound Radius** would be radius around the **Game Object** within which the
sound or music can be heard. Every microphone or player outside this radius will not
hear the sound or music.

The **Secondary End Event** would be the **Event** received by a **Game Object**,
when the sound or music had finished playing.

The priority of this **Event** would also determine the priority of the *stream*, relative to other *sound streams*. The priority of all *streams* would be described by a
single *Record*: the **PRIORITY END EVENTS RECORD**. This would have exactly
the same *Database Fields* as the **Priority Events Record**, used by the **Events Host**,
to determine the relative priority of **Secondary Events**. Except this one would only
contain the **Secondary End Events** of *sound streams*. And it would be used to set
the relative priorities of each one.

Each **Sound Microphone Object** would have a **SOUND MICROPHONE OBJECT RECORD**. The **Sound Microphone Object Record** would have the following *Fields:*

1. a *Primary Key* or **SOUND MICROPHONE ID**
2. **Object ID**
3. **MICROPHONE OFFSET X FIELD**
4. **MICROPHONE OFFSET Y FIELD**
5. **MICROPHONE OFFSET Z FIELD**

The **Sound Microphone ID** would be *Primary Key* of a **Sound Microphone Object Record**.

The **Object ID** would refer to the **Game Object** whose position in the *Game World* controls what can be heard by the player. And this would typically be the **Game Object** of the player's character. But it may be some other more remote **Game Objects** like a vehicle remotely controlled by the player.

The **Microphone Offset X, Y, Z Fields** is a relative position, around the microphone or **Game Object** from which the sound would be heard. This would be set depending on the type of game being played. For example, when you play a game from a third-person perspective, the **Camera Object** will be following and focused on the player's character. From a position which was slightly offset from the player's position. In that case, the microphone could also be offset by the same amount from the player's position.

During each *Unit of game time*, the **Sounds Host** would clear all the **Secondary End Events**, from the list of **Secondary Events** of the **Primary End Event** in its *Database Record.*

The **Sounds Host** would check the list of sounds which were being played in the **Sounds Played List Record**. If a sound or music had finished playing, then the **Sounds Host** would remove it from the list. The **Secondary End Event**, of that sound, would be added to the list of **Secondary Events**, in the *Database Record* of the **Primary End Event**.

After all the sounds being played had been similarly checked, to see whether these had finished, the **Sounds Host** would then send the **Primary End Event** to the **Events Host**. To signal the end of playback of all the sounds which had just ended.

The **Sounds Host** would then go through the list or queue of *sounds streams* that were meant to be played, in the **Sounds Waiting List Record**. If the **Game Object** for a *sound stream* were not within range of a **Sound Microphone Object Record**, then it would not be played. And it would be removed from the list. If the **Game Object** were within range of a microphone, then Primary Key of that *sound stream* would be moved, from the **Sounds Waiting List Record** to the **Sounds Playing List Record**.

Its *sound stream* would be loaded onto the computer hardware and played at the appropriate frequency, volume and on the *sound channel*, set in its **Sound Stream Record**.

When a new sound was added onto the list of sounds to be played, and its *sound channel* was occupied, the **Sounds Host** would find an empty *channel* to play it on.

And before playing that new sound, the **Host Module** would update the *Database Field*, that held its *sound channel*.

If all the *sound channels* were occupied, then the **Sounds Host** would find all the sounds being played with a lower priority than the new sound. The **Sounds Host** would stop the one with the lowest priority, add its **Secondary Event** to the list of the **Primary End Event** and send the **Primary End Event** to the **Events Host**. The new sound would then be played on the empty *sound channel*.

Any **Game Object** could directly add a sound to the **Sounds Waiting List Record**. But only a special **Game Object** or **MASTER SOUND SPEAKER OBJECT** should play sounds through that list. And all other **Sound Speaker Objects** that wanted to play a sound should send a **Secondary Event** to the **Master Sound Speaker Object**. Its properties would be held in a **MASTER SOUND SPEAKER OBJECT RECORD** with at least the following *Database Fields*:

1. a *Primary Key*
2. a **Game Object Code Field**

The **Master Sound Speaker Object** would use **SOUND SPEAKER SECONDARY EVENTS RECORD**. This would have the same *Database Fields* as **Secondary Events Records**. Except the **Game Object Field** would always be the same value i.e. the **Master Sound Speaker Object**. That is to say, this is a Database Table of Secondary Events which only the Master Sound Speaker Object receives. And there would be the following extra *Database Fields*:

1. a **Sound Stream ID**

The **Sound Stream ID** would be *Primary Key* of a **Sound Stream Record** containing the *sound streams* that the **Master Sound Speaker Object** would play when it received a **Secondary Event**.

The advantage of having all the sounds go through one **Game Object** and one *Database Table* is that you can see all the **Events** and sounds played in the game in one *Table*. And you can edit and control this in one *Table*.

There is a diagram showing the flow of information to and from the **Sounds Host** in Figure 3.11 with a Legend in Figure 3.12.

3.7   GAME CONTROLLERS HOST

The **Game Controllers Host** would read all *Game Controllers*. A *Game Controller* may contain any combination of *analogue devices*[33] and *digital devices*.[34] The *devices* would form one or more groups, on the *Game Controller*. But no single *device* would be in more than one group. For each group, the **Game Controllers Host** would send a **Connect Event** when the *Game Controller* was connected and a **Disconnect Event** when it was disconnected.

For *analogue devices*, the **Controllers Host** would send a **Controller Moved Event**, when these started to move. And it would send a **Controller Stopped Event**, when the *devices* stopped. For *digital devices*, it would send a **Controller Pressed**

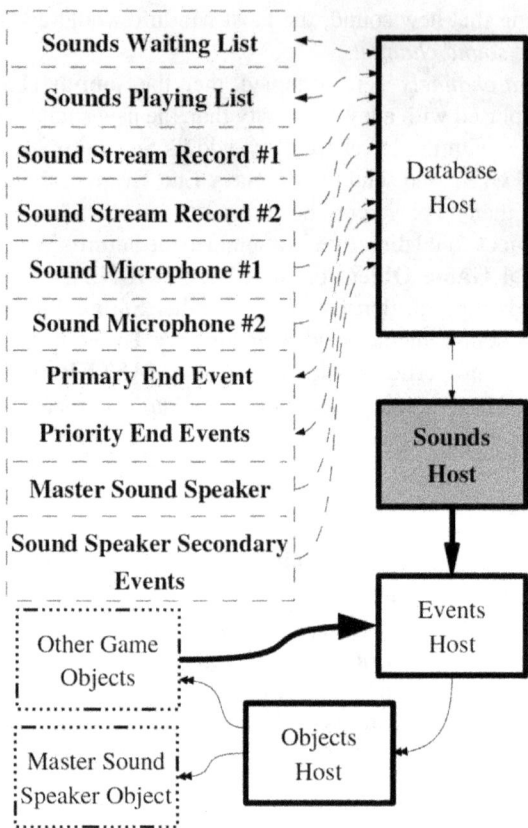

FIGURE 3.11   The flow of information to and from the **Sounds Host**.

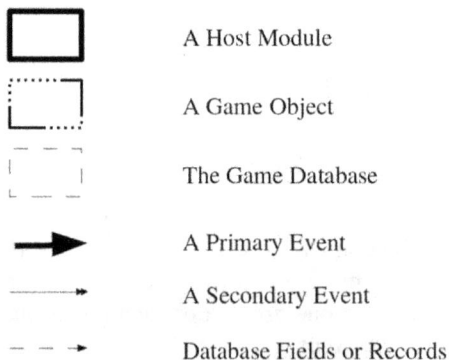

FIGURE 3.12   Legend of the symbols displayed in Figure 3.11. It is a list of the symbols, for the components of the **Event-Database Architecture**, that would interact with the **Sounds Host**.

Event when these were turned on. And it would send a **Controller Released Event** when these were turned off.

Each group of *devices*, or all the *devices*, on a *Game Controller* would have one *Record* in the **Game Database**. The details of each group would be held in this **DEVICE GROUP RECORD**. The *Software Developers*, building a game based on the **Event-Database Architecture**, would decide how many groups there would be for each type of *Game Controller*. They would also decide how many types of *Game Controller* they would permit, connected to the computer hardware.

The **Device Group Record** would have the following *Database Fields*:

1. a *Primary Key*
2. **DEVICE GROUP FIELD**
3. **CONTROLLER TYPE FIELD**
4. **Object IDs**

The **Device Group Field** would hold a list of unique words, that identified the *analogue devices* and *digital devices* in a group. Each *digital device*, or the axis of an *analogue device*, would be identified by one word. This word may be made up of one or more characters. And this would be used by the **Game Controllers Host** whenever it wanted to identify that *device*.

The **Controller Type Field** would hold the type of *Game Controller* these *devices* belonged to. For example, these could belong to a Keyboard, a Mouse, a Gamepad, a Touchpad, a Joystick etc.

The **Object IDs** would be the Primary Keys of the *Database Records* of the **Game Objects** that the *devices* had been assigned to. These **Game Objects** could either be **2D PLAYER OBJECTS** or **3D PLAYER OBJECTS** in 2D or 3D space respectively. The properties of these **Objects** would be held in either **2D PLAYER OBJECT RECORDS** or **3D PLAYER OBJECT RECORDS**. And each would have at least the following *Database Fields*:

1. a *Primary Key*
2. **Game Object Code Field**
3. *X Position*
4. *Y Position*
5. *Z Position* (in 3D space only)
6. *X Speed*
7. *Y Speed*
8. *Z Speed* (in 3D space only)
9. *X Acceleration*
10. *Y Acceleration*
11. *Z Acceleration* (in 3D space only)
12. **DEVICE MAPPING FIELD**
13. **CONTROLLER MAXIMUM FIELD**
14. **CONTROLLER CENTRAL FIELD**
15. **CONTROLLER MINIMUM FIELD**
16. **ANALOGUE HISTORY FIELD**

17. **ANALOGUE POSITIONS FIELD**
18. **DIGITAL HISTORY FIELD**
19. **DIGITAL POSITIONS FIELD**
20. **Secondary Connect Event**
21. **Secondary Disconnect Event**
22. **Secondary Controller Moved Event**
23. **Secondary Controller Stopped Event**
24. **Secondary Controller Pressed Event**
25. **Secondary Controller Released Event**.

Any movement of the *devices* would be applied to the properties of all the **2D Player Objects** or **3D Player Objects**, these had been assigned to. These **Objects** could either do nothing and just be used to store the activity of a *Game Controller*. Or these could do something more advanced, like control player characters in 2D or 3D *Game World*.

The **Device Mapping Field** would complement the **Device Group Field**. This would list the names of the numerical *Fields*, in the *Record* of the **Game Object**, which respectively would be changed by each *device* in the **Device Group Field**. The **Mapping Field** would be used to change the effect of the movement, of the *devices* on a *Game Controller*.

So, for example, rather than modifying the position, the movement of the *devices* could be easily redirected to the speed of a **Game Object** instead. This would be done by simply modifying the **Device Mapping Field**, from a list of the *Fields* for the

X Position,
Y Position and
Z Position

to a list of the *Fields* for the

X Speed,
Y Speed and
Z Speed.

When a *device* was moved, the movement would be applied to the *Fields*, of the **Game Object** the *device* had been assigned to. The number of *Fields* modified, would depend on whether it was a *digital device* or an *analogue device* such as a button or a *device* with many axes. If it were a *digital device*, a button or had only one axes, the *device* would only modify one *Field*. If it had two axes, the *device* would only modify two *Fields*. If it had three axes, it would modify three *Fields* and so on.

The *Fields* that would be modified would depend on the mapping of the *devices* listed in the **Device Group Field**, to the *Fields* listed in the **Device Mapping Field**. Each axis of an *analogue device*, in the **Device Group Field**, would have a corresponding entry in the **Device Mapping Field**. Each *digital device*, in the **Device Group Field**, would be paired with another. Both *devices* in each pair would appear consecutively in that *Field*. No two pairs would share the same *device*. Together, each pair would be treated as one *analogue device*, with one axes. So that when one was pressed, the *analogue device* would be at its highest point. And when the other was pressed, the *analogue device* would be at its lowest point.

Each pair of *digital devices* would have a pair of corresponding entries in the **Device Mapping Field**. These two numerical *Fields* would be modified when the *analogue device*, formed by the pair, was at its highest and lowest points respectively.

The amount these numerical *Fields* would be adjusted would depend on the **Controller Maximum**, **Controller Central** and **Controller Minimum Fields**. The **Controller Maximum Field** would hold the amount the numerical *Fields* should be moved, when a *device* was at the highest point of its axis. And the **Controller Minimum Field** would hold the amount the numerical *Fields* should be moved, when the *device* was at the lowest point. The amount of movement would be proportionally scaled between these two extremes.

For example, suppose the **Controller Maximum** and **Controller Minimum**, for a group of *analogue devices* and *digital devices*, were 10 and −10. And an *analogue device* were moved along an axis, from its default point, to 10% of the length between the default and the highest point. The numerical *Field* assigned to that axis would then be increased by 1. If the *device* were moved, subsequently, from 10% to 50% of this length, then its numerical *Field* would be increased by 4. Similarly, if the first of a pair of *digital devices* were pressed, then its numerical *Field* would be increased by 10. And if the second *device* were pressed, its numerical *Field* would be reduced by 10.

The **Controller Central Field** would hold the range of positions, about the default position of the *device*, within which any movement would be ignored. This would prevent any accidental movement, of an *analogue device*, from causing changes to the game.

The **Analogue History Field** would keep a history of the *analogue devices* that were used. It would initially contain an empty list. When each *analogue device* was moved or stopped along an axis, the unique word, for that axis, would be added onto the list. This list would have a fixed length. But it would be long enough to hold the word, for each axis in the group, three times. When the list was full, the oldest word, that had occurred more than twice in the list, would be removed. So that a new word could be added.

The **Analogue Positions Field** would also be used to keep a history of the *analogue devices*. But unlike the **Analogue History Field**, this *Field* would keep a history of the positions of the *analogue devices*. Like the **Analogue History Field**, it would initially contain an empty list. When each *analogue device* was moved or stopped along an axis, its position on that axis would be appended onto the list. The **Analogue Positions Field** would have the same fixed length as the **Analogue History Field**. And when an entry was removed from the **Analogue History Field**, its corresponding entry would be removed from the **Analogue Positions Field** as well.

Both that *Analogue History Field* and the *Analogue Positions Field* would determine when an *analogue device* had moved or stopped. These *Fields* would also determine how much further to adjust the numerical *Field*, which the *device* had been assigned to, when that *device* was moved.

Using the current position, and the last position of an *analogue device*, would determine when the *device* had moved. Using the current, the last and the last but one position, would determine when the *device* had just stopped. And by subtracting the last position, from the current position of the *device*, this would determine how much the *device* had moved. And this would be used to adjust the numerical *Field* that *device* had been assigned to. So that, as each *device* was moved about its default position, along an axis, each numerical *Field* would be adjusted proportionally about its initial value.

The **Digital History Field** would keep a history of the *digital devices*. It would initially contain an empty list. When each *digital device* was pressed or released, the unique word for that *device* would be appended onto the list. Just like the **Analogue History Field**, the **Digital History Field** would have a fixed length. But it would be, at least, long enough to hold the word, for each *digital device* in the group, three times. When the list was full, the oldest word, that had occurred more than twice in the list, would be removed. So that a new word could be added.

The **Digital Positions Field** would also be used to keep a history of the *digital devices*. However, unlike the **Digital History Field**, this *Field* would keep a list of the times that each *device* was pressed or released. Like the **Digital History Field**, it would initially contain an empty list. When each *digital device* was pressed or released, the *game time* would be appended onto the list. The **Digital Positions Field** would have the same fixed length as the **Digital History Field**. And when an entry was removed from the **Digital History Field**, its corresponding entry would be removed from the **Digital Positions Field** as well.

The **Secondary Connect**, **Disconnect**, **Controller Moved**, **Controller Stopped**, **Controller Pressed** and **Controller Released Event Fields** would hold **Secondary Events**. These **Events** would be received by the **Game Object** the group had been assigned to. These would be received when the group had been connected or disconnected, or one of the group had been moved, stopped, pressed or released respectively.

During each *Unit of game time*, if the *analogue devices* or *digital devices* in a group changed state, the **Game Controllers Host** would identify what **Event** had occurred. It would decide whether the group had been connected or disconnected. Or it would decide whether one of the group had been moved, stopped, pressed or released. It would prepare by clearing the list of all the **Secondary Events**, which were last sent for that **Primary Event**, from the *Record* of that **Event**.

For example, from the *Record* of the **Primary Connect Event**, it would remove all the **Secondary Connect Events** on that list. And, from the *Record* of the **Primary Disconnect Event**, it would remove all the **Secondary Disconnect Events** on that list and so on.

Game Controllers Host would send the new **Primary** and **Secondary Events** to the **Events Host**. And that would in turn send the **Secondary Events** to the **MASTER PLAYER OBJECT**. This is a special **Game Object** that all **Primary** and **Secondary Events** from the **Game Controller Host** would pass through. The **Master Player Object** in turn would apply the effects of the **Secondary Events** to the properties of **2D Player Objects** and **3D Player Objects**.

The **Master Player Object** would use the **MASTER PLAYER OBJECT RECORD** with the following *Database Fields*:

1. a *Primary Key*
2. a **Game Object Code Field**
3. **Secondary Connect Event**
4. **Secondary Disconnect Event**
5. **Secondary Controller Moved Event**
6. **Secondary Controller Stopped Event**
7. **Secondary Controller Pressed Event**
8. **Secondary Controller Released Event.**

If the group of *devices* had been connected, the **Game Controllers Host** would get the **Secondary Connect Event**, of the **Master Player Object**. And it would set the cause of that **Event** as the **Object** that that group had been assigned to in the *Database Table* of **Device Group Records**. And it would add this **Event** to the list of **Secondary Events**, of the **Primary Connect Event**. And after this had been done for all the groups of *devices*, the **Primary Connect Event** would be sent.

Similarly, if the group had been disconnected, the **Controllers Host** would get the **Secondary Disconnect Event**, of the **Master Player Object**. And it would set the cause of that **Event** as the **Object** that that group had been assigned to in the *Database Table* of **Device Group Records**. And it would add this **Event** to the list of **Secondary Events**, of the **Primary Disconnect Event**. And after this had been done for all the groups of *devices*, the **Primary Disconnect Event** would be sent.

The same steps would be followed when an *analogue device* or *digital device*, in the group, was moved, stopped, pressed or released. Each of these would result in the appropriate **Secondary Event** being taken from the **Master Player Object**. And the cause of that **Event** being set to the **Player Object** that group had been assigned to in the *Database Table* of **Device Group Records**. And the unique word that identifies the *analogue device* or *digital device* being added to the properties of that **Event**. And it would result in that **Event** being added to the *Record* of the corresponding **Primary Event**. And after this had been done for all the groups, that **Primary Event** would be sent.

When the **Master Player Object** receives a **Secondary Event**, it forwards this on to each of the **Game Objects** which are listed as the causes of that **Secondary Event**. So long as each **Object** is not the **Master Player Object** itself. And when it forwards these **Events** on to those **Game Objects**, it changes the cause of that **Event** to itself.

When *analogue devices* were moved or stopped, the **Game Controllers Host** would measure how much *each device* had moved. And it would add this amount and the unique word for this axis to the properties of the **Secondary Event** sent to the **Master Player Object**. And the **Master Player Object** would use the amount to adjust the numerical *Fields*, assigned to the axis along which the *devices* had moved, proportionally. It would then add the unique words, for this axis, onto the **Analogue History Field**. And it would add the new position of the *device*, along this axis, onto the **Analogue Positions Field**.

When *digital devices* were pressed or released, the **Game Controllers Host** would decide which *device* had changed state. It would add the unique word for this *device* to the properties of the **Secondary Event** sent to the **Master Player Object**. And the **Master Player Object** would then adjust the numerical *Fields* assigned to that *device*. It would add the unique words, for each *device*, onto the **Digital History Field**. And it would append the *game times*, at which the *device* was used, onto the **Digital Positions Field**.

The **Master Player Object** would look at the **Digital History Field** and **Analogue History Field** of all the **2D** and **3D Players Objects**. It would look for each single word for a single *device*, or sequence of words for a group of *devices*, in the history which constituted a command or action to be performed by the player's character. The single word or sequence of words which constituted a command would be defined in a **DEVICE SEQUENCE PRIMARY EVENTS RECORD**. This would map each single word or sequence of words to a **Primary Event** for a command. The *Record* would include the following *Database Fields*:

1. a *Primary Key*
2. a sequence of one or more *digital devices* or *analogue devices* that constituted a command
3. the **Primary Event** that should be sent when that sequence was detected in the history of devices of a **2D** or **3D Player Object**.

When the **Master Player Object** detected a word or sequence of words for a command in the **Digital History Field** or **Analogue History Field** of a **2D** or **3D Player Object**, it would send the corresponding **Primary Event**. And whatever **Game Objects** had tied its **Secondary Events** to that **Primary Event** would respond with whatever **Actions** these deemed appropriate. The **Master Player Object** would then clear the sequence of words for that command just executed from the history.

The structure of the *Database Records* of the **Primary Events** sent by the **Game Controllers Host** i.e.

- **PRIMARY CONNECT EVENT RECORD**
- **PRIMARY DISCONNECT EVENT RECORD**
- **PRIMARY CONTROLLER MOVED EVENT RECORD**
- **PRIMARY CONTROLLER STOPPED EVENT RECORD**
- **PRIMARY CONTROLLER PRESSED EVENT RECORD**
- **PRIMARY CONTROLLER RELEASED EVENT RECORD**

would be the same as other **Primary Events**. The structure of the *Database Records* for the **Secondary Events** sent by the **Game Controllers Host** i.e.

- **SECONDARY CONNECT EVENT RECORD**
- **SECONDARY DISCONNECT EVENT RECORD**
- **SECONDARY CONTROLLER MOVED EVENT RECORD**
- **SECONDARY CONTROLLER STOPPED EVENT RECORD**
- **SECONDARY CONTROLLER PRESSED EVENT RECORD**
- **SECONDARY CONTROLLER RELEASED EVENT RECORD**

would be the same as other **Secondary Events**. Except it would include the following additional *Database Fields*:

1. the unique word for an *analogue device* or *digital device* of the *Game Controller* whose change in position along an axis or change in state caused this **Secondary Event**
2. the amount of change that occurred

The **Game Controllers Host** would be the last of the **Host modules**, which the **Event-Database Architecture** would use to implement a *game design*.

There is a diagram showing the flow of information to and from the **Game Controllers Host** in Figure 3.13 with a Legend in Figure 3.14.

3.8   CENTRAL HOST

To operate, all the seven **Host modules** would be synchronised by a **Central Host**. This would start the whole game. It would set up any *software libraries*, the **Host**

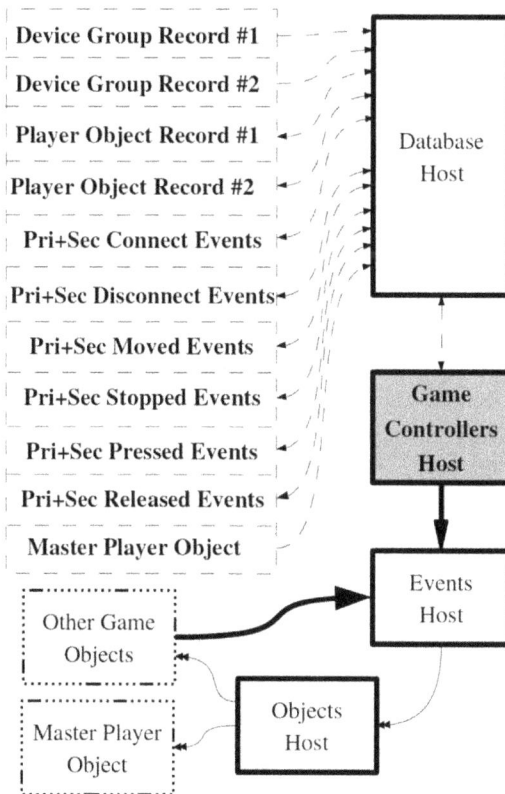

FIGURE 3.13   The flow of information to and from the **Game Controllers Host**.

A Host Module

A Game Object

The Game Database

A Primary Event

A Secondary Event

Database Fields or Records

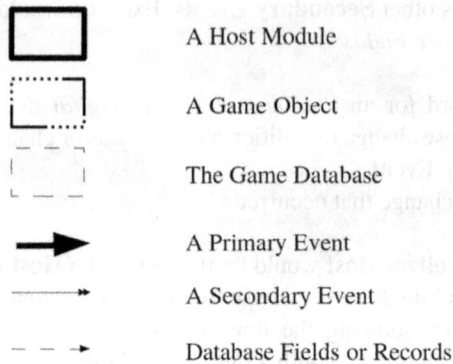

FIGURE 3.14   Legend of the symbols displayed in Figure 3.13. It is a list of the symbols, for the components of the **Event-Database Architecture**, that would interact with the **Game Controllers Host**.

Modules and the **Game Objects** would use. It would also set up any computer hardware these would use. After the software and hardware had been prepared, it would start and setup all the other **Host Modules**, beginning with the **Database Host**. It would send the **Primary Initial Reset Event**, to the **Events Host** after all the other **Host Modules** had been set up successfully.

If an error occurred with a **Host Module**, while it was setting up, then it would indicate this by sending a *Primary Key* to the **Central Host**. All known errors that could occur, in the **Host Modules** and **Game Objects**, would have an **ERROR RECORD**. This would have the following *Database Fields*:

1. a *Primary Key*
2. a text describing an error

The **Central Host** would use the Key it received, to look up the **Error Record**, with the text describing the error, display the text, and log it in a computer file next to the **Game Database**.

If this were not possible because, for example, the **Database Host** was not set up yet, then **Central Host** would simply display and log the *Primary Key*. Either way, it would wait an indefinite amount of time for the error to be read. After that, the **Central Host** would shut down each *module*, shut down the *software libraries* and hardware and close the game.

With the exception of the **Database Host** and the **Objects Host**, the rest of the **Host Modules** would then be managed by the **Central Host**, once the game started successfully. It would periodically update each of these *modules* once, during each *Unit of game time*. By sending it how much time had passed since the game started. After each attempt, the **Central Host** would check whether any errors had occurred when the *module* was updating its task, causing it to fail. The manner in which it would detect these errors, and respond, would be the same as when it setup all the *modules*.

If all *modules* were updated successfully, before the *Unit time* had elapsed, then the **Central Host** would wait for the next period to begin. Any reasonable amount of time could be chosen as the *Unit time*. But typically, most *Software Developers* would choose a time dependent on how fast they wanted the computer screen to be refreshed. This means the time between each *Frame* of the game being displayed.

During each period, the **Physics Host** would be updated first, followed by the **Game Controllers Host** and the **Events Host**. The **Sounds Host** and the **Graphics Host** would be updated last, in that order.

The **Database Host** would obviously only perform its task when the **Game Database** was accessed. The **Objects Host** would only perform its task when it received a **Secondary Event**.

Apart from **Error Records**, the **Central Host** would use one other *Record* in the **Game Database**. This would be the **GAME TIME RECORD**. This would contain the following *Database Fields*:

1. a *Primary Key*
2. the *Unit of game time* being used, when the game started
3. how much time had elapsed since the start.

The latter of these would be updated, by the **Central Host**, at the end of each *Unit of game time*.

Finally, as previously explained, when the **Events Host** received a **Primary Shutdown Event**, it would shut itself down. When the **Central Host** tries to contact the **Events Host** it should get no response, and it should treat this as any other error. That is to say, it should display and log the text describing that error. Or if no text were available, it should display the *Primary Key* for that error on the screen or write it in the logs. After that it would shut down the remaining *modules*, shut down the *software libraries* and hardware and close the game.

There is a diagram showing the flow of information to and from the **Central Host** in Figure 3.15 with a Legend in Figure 3.16.

3.9   THE NETWORK OF THE ARCHITECTURE

All the main seven **Host Modules** would either send information to, or receive it from, the **Events Host** or the **Database Host**. Most of the **Host Modules** would have an *Interface* with both.

The **Database Host** would play a central part. All of the **Host Modules** would either send or receive information from it. This would give *Game Producers*, *Designers* and other staff control over the whole **Architecture** through the *Database*.

All of the **Game Objects**, and some of the **Host Modules**, would send information to the **Events Host**, but only the **Objects Host** would receive information from it. This is consistent with the principles of the **Architecture**, as outlined in the initial

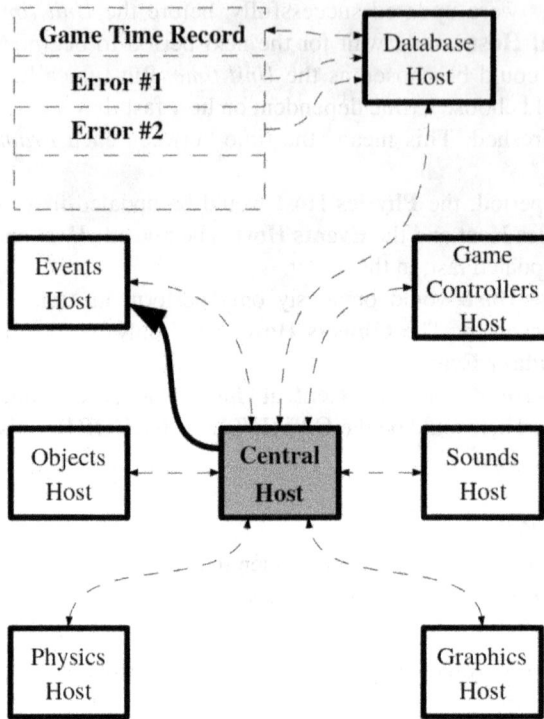

FIGURE 3.15   The flow of information to and from the **Central Host**.

description of the solution. **Events** would only be a means of controlling the flow of the game; not the operation of the **Architecture** itself.

There is a diagram showing the flow of information to and from the **Host Modules** in Figure 3.17 with a Legend in Figure 3.18. There is a table describing this information and its origins in Table 3.1.

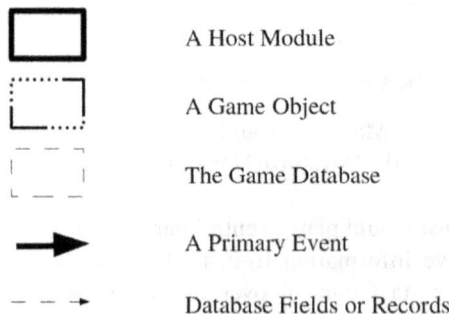

▭	A Host Module
⌐⌐	A Game Object
⌐⌐	The Game Database
➤	A Primary Event
- - ➤	Database Fields or Records

FIGURE 3.16   Legend of the symbols displayed in Figure 3.15. It is a list of the symbols, for the components of the **Event-Database Architecture**, that would interact with the **Central Host**.

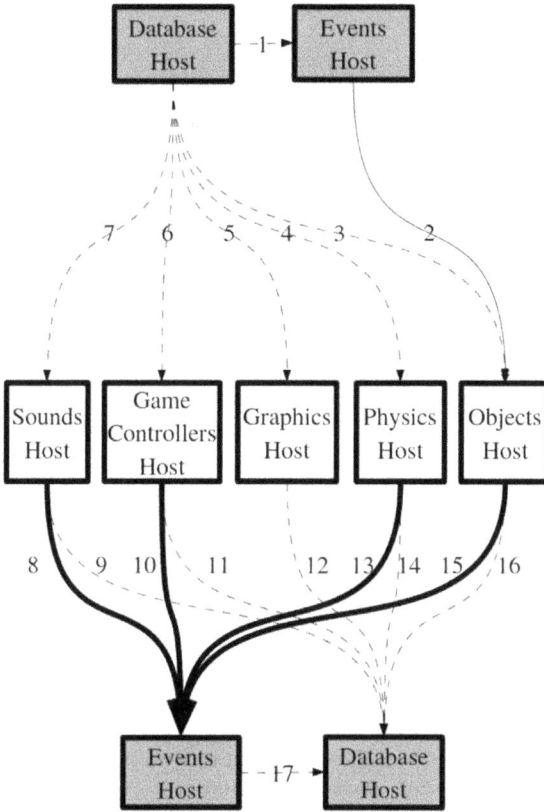

FIGURE 3.17 The flow of information between the two core **Host Modules** of an **Event-Database Architecture** and the peripheral **Modules** and back again.

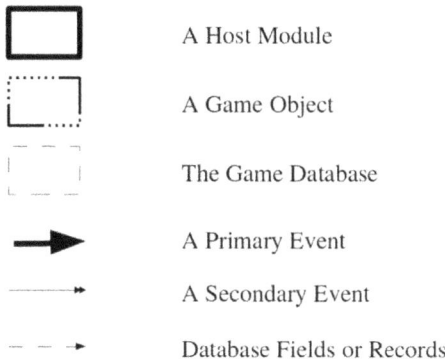

FIGURE 3.18 Legend of the symbols displayed in Figure 3.17. This includes the symbols for the **Host Modules**, **Game Objects**, **Game Database**, **Primary Events**, **Secondary Events**, *Database Fields* and *Database Records* of the **Event-Database Architecture**.

TABLE 3.1
Legend of the Numbers Displayed in Figure 3.17

Data	Role
1	List of **Game Objects** that would receive each **Secondary Event**. The properties of each **Event** (e.g. The delay before the **Event** was sent).
2	Customised **Secondary Events** for a particular game (e.g. A Reset Event, a Collision Event, a Move Event etc.).
3	Properties of **Game Objects**, *Abstract data* or shared *Game data*.
4	List of **Game Objects** whose physical properties should be updated. The properties of these **Objects** (e.g. position, speed, acceleration, orientation, bounding shape etc.).
5	List of cameras whose view of the *Game World* should be displayed. List of **Game Objects** that should be displayed. The properties of these **Objects** (e.g. *Vertices*, *Textures*, *Texture coordinates* etc.).
6	The **Game Object** to update when each *Game Controller* was manipulated (e.g. pressed, released, moved etc.).
7	List of sounds to be played or were being played. The properties of these sounds (e.g. *sound channel*, volume, frequency, encoded *sound stream* etc.).
8	**Primary End Event** when one or more sounds had finished playing.
9	**Secondary End Events** of each sound when it had finished playing. Updated list of sounds being played.
10	**Primary Events** sent when the *Game Controllers* were manipulated (e.g. pressed, released, moved etc.).
11	Updated properties of **Game Objects** that have been assigned to a **Game Controller** (e.g. new positions, **Digital Group Fields** etc.).
12	Projection of 2D or 3D shapes, through the 2D or 3D cameras. Updated list of projections that would be displayed.
13	**Primary Proximity** and **Collision Events**.
14	Updated physical properties of **Game Objects** (e.g. position, speed, acceleration, rotation etc.).
15	Customised **Primary Events** for a particular game.
16	Updated properties of **Game Objects**, *Abstract data* or shared *Game data*.
17	Updated properties of delayed **Secondary Events** and the list of delayed **Events**.

List of examples of the information exchanged between the *Modules* of an **Event-Database Architecture**.

There is a diagram showing the flow of information between the staff and the **Game Database** in Figure 3.19 with a Legend in Figure 3.20. There is a table describing this information and its origins in Table 3.2.

3.9.1 SINGLE USER MONOLITHIC FORM

When you run or start your computer, it starts an *Operating System*.[35] This is software which allows other software that you use or *Software*

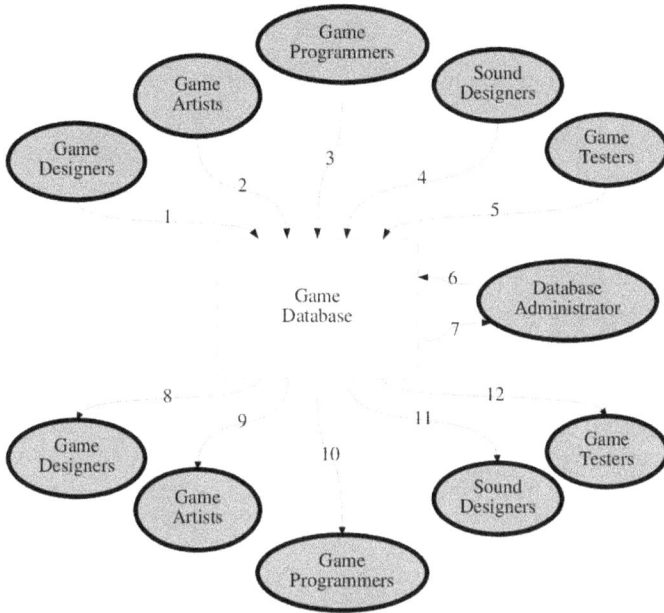

FIGURE 3.19 The flow of information between the staff using the **Event-Database Architecture**.

Applications[36] to share the same resource. In this case, that resource is the computer hardware i.e.

Mouse
Keyboard
Game Controllers
Screen
Storage media or Hard Disk
Computer memory
Peripherals connected such as a Printer
Network Cards
Graphics Cards
Sound Cards and so on.

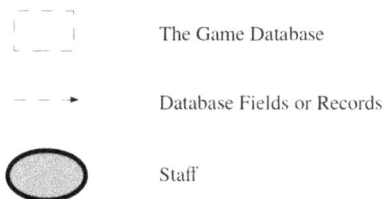

The Game Database

Database Fields or Records

Staff

FIGURE 3.20 Legend of the symbols displayed in Figure 3.19. This includes the symbols for the staff and the **Game Database**.

TABLE 3.2
Legend of the Numbers Displayed in Figure 3.19

Data	Role
1	Modified chain of **Events** that would meet the latest requirements of the *game design*. New **Game Objects** that should be added to the *Game World*. Old **Game Objects** that should be removed. New order of **Game Objects** (i.e. the stage each one would appear). Modified properties of **Game Objects** (e.g. new size, position, appearance etc.).
2	New graphics and animation of **Game Objects**. New fonts for displaying texts. Combinations of graphics (e.g. lighting of **Game Objects**, the tiling of *Textures*, the overlaying of *Textures* etc.). Animation of *Textures*. New properties of graphics (e.g. *Textures*, overlaying colours, shading, tone, graphical effects etc.).
3	New **Game Objects** that should be added to the *Game World*. Old **Game Objects** that should be removed. New and modified properties of **Game Objects**. New **Events** and **Actions** (i.e. **Secondary Events**) for **Game Objects**. Modified chain of **Events** that would meet the latest requirements of the *game design*.
4	New music or sound effects for **Game Objects**. Combinations of sounds (e.g. which sounds would play together on each stage of the game, which sounds would overlap, which sounds would follow one another etc.). New and modified properties of *sound streams* (e.g. volumes, *sound channels*, reverberation, other effects etc.). New **Secondary Events** to play *sounds streams*. Modified chain of **Events** to play *sound streams*.
5	Modified chain of **Events** that would meet the requirements of the test plan. Modified properties of **Game Object** to meet the requirements of test plan. Modified properties of **Database Monitor Record**, to monitor the properties of **Game Objects** and other *Records* during the execution of test plan.
6	Corrections to entries in the *Database Tables, Records* or *Fields*. Redundant *Database Tables, Records* or *Fields* that should be removed. New order of *Records* or *Fields* that would improve the performance of the *Database*. Changes to the *data* which each **Game Object** or staff could access. New names for *Database Tables, Records* or *Fields*.
7	Names of the various items or entities that were stored in the **Game Database** (hence the words of the language of the project). Current order of all *Database Tables, Records* or *Fields*. Overview of all the entries in the *Database Tables, Records* or *Fields* that may be corrected.
8	Overview of all the chain of **Events** that may be changed. Overview of all the **Game Objects** that may be added or removed. Overview of all the properties of these **Game Objects**. Overview of the order in which all the **Game Objects** would appear in the game.
9	Overview of all the graphics and animations for **Game Objects**. Overview of all combinations of the graphics. Overview of all the animation of the graphics. Overview of all the properties of graphics.
10	Overview of all the **Game Objects**. Overview of all the chain of **Events** that may be changed. Overview of all the **Secondary Events** that **Game Objects** would respond to and hence their **Actions**. Overview of the properties of all **Game Objects**.
11	Overview of all the music and sound effects for **Game Objects**. Overview of all the combinations of *sound streams*. Overview of all the properties of *sound streams*. Overview of **Secondary Events** that play *sound streams*. Overview of chain of **Events** that play *sound streams*.
12	Overview of all the individual **Events** and **chain of Events** that could be tested. Overview of all the properties of **Game Objects** that could be tested. Overview of all the **Database Log Records** containing the *data* modified during the execution of the test plan.

List of examples of the information exchanged with the **Game Database**, by the staff. So that they could control the game and its production.

An example of a *Software Application* is a Word Processor, a Web Browser, a Clock, a File Manager or a Computer Game built with the **Event-Database Architecture**. A *Software Application* can come in many forms.

In its most simplest form a *Software Application* is contained in one file on the storage media or Hard Disk of the computer. And when you run or start that *Software Application* it starts one *Process*[37] in the *Operating System*, using that one file. And when that *Software Application* ends, or is shut down, the *Process* stops. This is the monolithic form or **SINGLE USER MONOLITHIC FORM** of a *Software Application*.

In a slightly more complex form a *Software Application* is contained in two or more files. And when you run or start the *Software Application* it starts two or more *Processes*, in the *Operating System*, using those files, one *Process* from each file. Each *Process* has its own independent space in computer memory. These *Processes* run in parallel and communicate with each other in order act as one system. Sometimes these *Processes* share the same space in computer memory and are called *Threads*.[38] The Software User is not aware of these parallel *Processes* or *Threads*. As far as the User is concerned the *Application* acts as one system. And when that *Software Application* ends, or is shut down, all of the *Processes* or *Threads* stop. This is the multi-Threaded form or **SINGLE USER MULTI-THREADED FORM** of a *Software Application*.

A variation of this **Multi-Threaded Form** is another more complex form where the *Software Application* again is contained within two or more files. But the files are copied and distributed across two or more computers on a computer network. Since the files are distributed on different computers on the network, you either require two or more Software Users to synchronize starting the *Software Application* on each computer. Or one User who can start the *Software Application* on two or more computers at the same time. Again, when they run or start the *Software Application*, it starts two or more *Processes* on multiple *Operating Systems*, on multiple computers on the network. These *Processes* run in parallel and communicate with each other, across the computer network, in order to act as one system. And when that *Software Application* ends, or is shut down, all of the *Processes* running on the computer network stop. This is the distributed form or **MULTI-USER DISTRIBUTED FORM** of a *Software Application*.

With the **Event-Database Architecture** you can create a Computer Game in either the **Monolithic**, **Multi-Threaded** or **Distributed Form**. The **Form** you would use would depend on what type of game you are creating and the computer hardware. Some computer hardware do not have *Operating Systems*. Some *Operating Systems* do not allow multiple *Processes* or *Threads* to run in parallel. And depending on the **Form** you use, there may be additional requirements of the **Host Modules**.

If you were creating a single-player game then you would create it with the **Event-Database Architecture**, in either a **Monolithic Form** or a **Multi-Threaded Form**.

In a **Monolithic Form**, one file would contain all of code for the eight **Host Modules** of the **Architecture**. The advantage of this form is that it is the simplest and would not need any additional requirements of the **Host Modules**. And the communication between the **Host Modules** will be very fast. The disadvantage of this form is that if any one **Host Module** had an error or failed, then the whole

Architecture would fail. And the game would shut down. You would need to fix the code, rebuild the game and restart it to continue playing.

In a **Multi-Threaded Form**, however, the **Host Modules** would require additional features. Namely, each **Host Module** would also be required to be able to find and communicate with other **Host Modules**.

For example, in a **Multi-Threaded Form**, you could decide that each **Host Module** will be in a separate file and run in a separate *Process* with its own independent space in the computer memory. In this case you would need eight files for the eight **Host Modules**. And you would need another file, in a set location in the File system on the storage media or Hard Disk of the computer, that all the **Host Modules** have access to. Through which the **Host Modules** could find and communicate with each other. The advantage of this method is that if any one **Host Module** had an error or failed, then the whole **Architecture** would not fail. And you can rebuild the **Host Module** that failed or restart it, to resume playing the game. The disadvantage of this method is that it is more complex. And the common file which the **Host Modules** share access to, and use to communicate, could be a bottleneck and a source of problems. If you have many *Processes* trying to access it at the same time.

Alternatively, you could decide that each **Host Module** will run in a *Thread*, sharing the same space in computer memory as other **Host Modules** running in other *Threads*. The advantage of this method is that you do not need a common file that the **Host Modules** share access to, to find and communicate with each other. Therefore that file will not be a bottleneck or a source of problems.

The disadvantage of this method is that it is more complex. If any one **Host Modules** failed, then the whole **Architecture** would fail. And you would need to rebuild the **Host Modules** and restart the game. Another disadvantage is that each **Module** has to be aware that it is sharing the same space in computer memory with other **Host Modules**. So it should only read or write *data* in computer memory if other *Threads* were not using that *data*. This is usually achieved by locking access to that *data*, when reading or writing to it. And that in turn can lead to situations called deadlocks, where all the *Threads* or **Host Modules** are waiting for some other *Thread* to release access to some *data*.

3.9.2   Multi-User Distributed Client Server Form

If you want to create a multi-player game, then you have to build a game with the **Event-Database Architecture** in a **Distributed Form**.

In a **Distributed Form**, Each of the eight **Host Modules** would be contained in one file. And each file would start one of the **Host Modules** in a separate *Process* in the *Operating System*. And the files would be duplicated and distributed across multiple *Operating Systems* on multiple computers on the local computer network in one of two ways. Depending on whether you want one large powerful central computer doing all the work and handling the majority of the *Processes*. And the rest of the *Processes* being handled by smaller less powerful computers. Or whether you want to distribute the work and *Processes* evenly across the computers on the network. The former way is called a **CLIENT SERVER NETWORK ARCHITECTURE**. And the latter way is a called a **PEER TO PEER NETWORK ARCHITECTURE**.

In a **Client Server Network Architecture**, the large powerful central computer doing the majority of the work, and handling the majority of the *Processes* is called the Server or in Computer Games a **GAME SERVER**. This would run the

Database Host
Events Host
Objects Host
Physics Host
Central Host

of the **Event-Database Architecture**.

The smaller less powerful computers doing the rest of the work, and running the rest of the *Processes* are called the Clients, or in Computer Games the **GAME CLIENTS**. Each player in a multi-player game plays the game through one **Client**. Each **Client** would run its own instance of the

Graphics Host
Game Controller Host
Sound Host

When started, each of these **Host Modules** would connect to one or both *Processes* on the **Server**, running the **Events Host** and the **Database Host**.

When the **Graphics Host** was started on the **Client** it would connect to the **Database Host** on the **Server**. And start getting information about what to render on the screen of the **Client**.

When the **Game Controller Host** was started on the **Client**, it would connect to the **Events Host** on the **Server**. And send **Primary Events** to the **Events Host** as and when the buttons or axis on the Mouse, Keyboard or *Game Controllers* on the **Client** were used.

When the **Sounds Host** was started on the **Client**, it would connect to the **Event Host** and **Database Host** on the **Server**. And start getting information about what sounds to play on the **Client** from the **Database Host**. As well as what **Primary Events** to send to the **Events Host**.

With this **Distributed Form** of the **Event-Database Architecture** there will be an additional requirement. Each **Host Module** will need to be able to connect and send messages to other **Host Modules** running in *Processes* in *Operating Systems* on other computers on the local computer network. Each **Host Module** will need to be able to listen for connections from other **Host Modules**.

When you start each **Host Module** you should be able to either specify the address of the other **Host Modules**, on the computers on the local computer network. That you want to make an outgoing connection to. Or you should be able to specify the address on the local computer that the **Host Module** should listen for incoming connections.

When you specify the address of a *Process* running on a computer you normally specify two numbers called a *TCP/IP Address*[39] and a *Port Number*.[40] Therefore when you start the **Host Modules** it should allow you to specify these two numbers and whether you want the **Host Module** to make an outgoing connection or listen for incoming connections.

But you do not have to specify this for each **Host Module** manually. You just need to start the *Software Application*, and specify these two numbers, and the incoming or outgoing parameter. The *Software Application* should automatically pass this on to the **Host Modules** when it starts the *Process* running each **Host Module**.

In the **Event-Database Architecture**, the **Central Host** is the one that starts the *Software Application*. It is the one that starts all the other **Host Modules**. So in a **Distributed Form** of the **Architecture**, the **Central Host** would need to be the one that would recognise those three parameters i.e.

1. *TCP/IP Address*
2. *Port Number*
3. outgoing or incoming

However, there can only be one instance of the **Central Host** in the **Architecture**. Therefore if a **Central Host** was started with these two parameters and the 'outgoing' parameter, it should assume that there is another instance of the **Central Host** on the computer network. And after starting the other **Host Modules** with the parameters to direct them to that other computer, the **Server**, the **Central Host** would shut down. But the other local *Processes* in the local *Operating System* running the other **Host Modules** on the **Client** would continue.

If the **Central Host** was started with the 'incoming' parameter, then that would indicate that it was being started on the **Server**. And therefore it should not shut down but continue to perform its function and control all of the other **Host Modules** that make an incoming connection to the **Central Host**.

There are two further additional requirements for the **Graphics Host**, **Game Controller Host** and the **Sounds Host** when the **Event-Database Architecture** is used to create a multi-player game. These requirements come from the fact that you need to render multiple views of the *Game World* to multiple players or Software Users on the computer network. And you need to play back sounds from the *Game World* from multiple viewpoints. And you need to identify and authenticate the **Game Controllers** on the different computers on the computer network being used to play the game. To ensure that there was no cheating.

Beginning with the **Game Controller Host**, and the **Device Group Record** it uses, and the **Device Group Field** in that *Record*, this *Field* has to include information for authentication. This includes authenticating the **Game Controllers** and players or Software Users on the network. In a single player game, the **Device Group Field** would contain a unique word identify a unique local device that the player can use the control the game e.g. Joystick1.

But in a multi-player game, this **Field** should also contain one or more of these pieces of information, separated by a colon

1. *TCP/IP Address*
2. *Username*[41]
3. *Password*[42]
4. *Authentication Token*[43]

e.g.   Joystick1:192.168.0.1:Player1:Password1234:54&QA>65R,3I087-S=V]
R9#$R,S0*.

The *TCP/IP Address* is a word that identifies the **Client** or computer being used
by the player to play the game e.g. 192.168.0.1.

The *Username* is a name identifying a player. This can be their real name or some
pseudo name e.g. Player1.

The *Password* is a word that is encrypted or decrypted from a keyword that only
the player knows and always uses to log on to the game e.g. Password1234.

The *Authentication Token* is a word that is encrypted from a compound of the
Username and *Password*, that only the *Software Application* knows how to encrypt
and decrypt e.g. UlF1YXllOlN1bW9EaWdpdGFsSmVkaTIwMjU=

In a single player game, the **Device Group Field** only resides in the **Device
Group Record** of the **Game Controllers Host**. But in a multi-player game the
Device Group Field would be added to the **Camera Object Record** used by the
Graphics Host, and the **Sound Microphone Object Record** used by the **Sounds
Host**. To identify which **Camera Object** and microphone would be owned by which
Client or player on the computer network. And to show only the **Game Objects** that
can be seen by that **Camera Object** or play the sounds that can be heard by that
microphone, owned by that **Client** or player.

When you start the **Host Modules** on the **Client**, i.e. the **Graphics Host**, **Game
Controllers Host** and the **Sounds Host**, you would pass either these three pieces of
information in the parameters i.e.

1. *TCP/IP Address*
2. *Username*
3. *Password*

Or you would pass these two pieces of information in the parameters i.e.

1. *TCP/IP Address*
2. *Authentication Token*

And the **Host Modules** would in turn pass it on to the **Events Host** and the
Database Host on the **Server** when a connection was made. And the **Host Modules**
on the **Server** would search the **Game Database** for the **Device Group Record**,
Camera Record or **Sound Microphone Object Record** that contained these pieces
of information in its **Device Group Field**.

If a matching *Field* were not found, then the connection would be denied. If a
matching *Field* were found, then the connection would be accepted. And the remote
Graphics Host would identify the **Camera Record** it should use to display the
Game World to the player with this **Device Group Field**. And the remote **Game
Controllers Host** would identify the **Device Group Record** it should use to control
the player's character with the same **Device Group Field**. And the **Sounds Host**
would identify the **Microphone Object Record** it should use to play back sounds to
the player with the same **Device Group Field**.

Whether you choose to include all four pieces of information, three, two or just one in the **Device Group Field**, will depend on the level of security you want. It may be beneficial, while a game was being developed, to have a low level of security. And just use the *TCP/IP Address* of the **Client** for authentication. And later on during the production process, when a game was about to be released, switch to a high level of security and use all four. Since these four pieces of information will have to be specified every time in the parameters used to start the **Host Modules** on the **Clients**.

One advantage of this system is that two players using two **Clients** on the computer network, can connect to the **Server** with a **Game Controller Hosts**, using the same authentication information, and control the same **Device Group Record**. And that in turn means they can control the same **Player Object Record** referred to by the **Device Group Record**. And that in turn means they can control the same **Player Object** or player's character in the *Game World*.

This may seem counterintuitive at first. You may think that only one player should be able to control one **Player Object** in the *Game World*. But imagine a game where the **Player Object** was a large complex vehicle, which was so complex that it required multiple players to control it. Like a space ship, or naval ship, or a large robot. In that case having two players, on two **Clients** on the network, control the same **Player Object** would make sense.

Another advantage of this system is that two players using two **Clients**, can connect to the **Server** with a **Graphics Host**, with the same authentication information and use the same **Camera Object Record**. And that in turn means that the two **Clients** or players will see the same *Game World* on their screen.

Again this may seem counterintuitive at first. You may think that only one player should be able to see the view of the *Game World* through one player's character or Game Camera. But imagine a game where you want other players to be able to spectate and see the *Game World* through eyes of a player participating in the game. Many commercial *game-engines* already have this feature.

But with these *game-engines* you have to explicitly add a new **Game Object** which will act as a spectator. And the process for doing this is as long and as complex as adding a **Game Object** for a normal player, including writing new code. Whereas with the **Event-Database Architecture** you can implicitly spectate the game, without having to add a new **Game Object** or write new code.

Another advantage of this system is that one player using two **Clients** can connect to the **Server** with a **Sounds Host**, and with two different authentication information, that use two different **Sound Microphone Object Records**. And that in turn means the two **Clients** will hear the *Game World* from two viewpoints. If the two **Sound Microphone Object Records** were attached to the same **Player Object**, but with different offsets around that player, then that player will hear true stereophonic sound coming from the two **Clients**. As the player's character moved through the *Game World*. And you can use this technique to increase the number of microphones around the player's character, from two, to four, to five. At which point the player will be hearing true quadraphonic or surround sound, from the four or five **Clients**.

3.9.3 MULTI-USER DISTRIBUTED PEER TO PEER FORM

As explained in the previous chapter, a **Client Server Network Architecture** is very similar to a **Peer to Peer Network Architecture**. The difference is that in a **Client Server Network Architecture** the majority of the *Processes* doing the work, of a *Software Application*, are on one large central computer called the **Server**. And the rest of the *Processes* doing the lesser work are on multiple less powerful computers called the **Clients**. In a **Peer to Peer Network Architecture**, the *Processes* are all distributed evenly across all the computers on the local computer network which are equally as powerful. Each of these computers is called a Peer or in Computer Games a **GAME PEER.** Each player plays the game through one **Peer**.

So likewise, the **Distributed Form** of the **Event-Database Architecture** in a **Peer to Peer Network Architecture** is very similar to the **Form** for a **Client Server Network Architecture**. And the additional requirements of the **Host Modules** is very similar to the requirements for a **Client Server Network Architecture**. There are five major differences.

The first major difference is that instead of one computer, the **Server**, running the *Processes* of five **Host Modules** and the **Clients** running the *Processes* of three **Host Modules**, in the **Peer to Peer Network Architecture**, all the **Peers** would be running the *Processes* of seven **Host Modules**. Each computer would be running its own instance of

Database Host
Events Host
Objects Host
Physics Host
Graphics Host
Game Controller Host
Sounds Host

The second major difference is that one special **Peer** would be running in addition to these seven **Host Modules** its own permanent instance of the **Central Host.**

This would be one and only instance of the **Central Host** on the computer network. That would synchronise the operation of all the other **Host Modules** on the network. All the other **Peers** would have a temporary **Central Host**, that would only be used to start the other **Host Modules** on that **Peer.** With the parameters required to connect them to the permanent **Central Host** on the special **Peer**. But after that this temporary **Central Host** would shut down. Leaving only the *Process* of the permanent **Central Host** on the special **Peer** running.

The third major difference is that when started each **Host Module** on each **Peer** would connect to the instance of the **Database Host** and **Events Host** on the local computer. Instead of connecting to the instances on a remote computer.

The fourth major difference is that each **Host Module** would also connect to the permanent **Central Host** on the special **Peer**. That would send messages to the **Host Modules** on the network to synchronise their operation.

The fifth major difference is that when each instance of the **Events Host** received an **Event**, it would send it to the **Central Host** on the special **Peer**. And that **Central Host** would send that **Event** to all other instances of **Events Host** on all the other **Peers**. So that the **Event** would be replicated on all the **Peers** on the computer network. And thus synchronise the *Game Worlds* being modelled on all of the **Peers**.

There are at least two problems with this system that can affect the synchronisation of the *Game Worlds*. The first problem is that the effect of an **Event** may not be deterministic. That is to say, an **Event** being received by an instance of the **Events Host**, on one computer, may not necessarily produce the same effect on the **Game Database** as the same **Event** being received by another instance of the **Events Host** on another computer. It could be, for example, that **Event** causes a random number to be added to a *Field* or a random *Database Record* in the **Game Database** to be modified. Or it could be that **Event** causes a random **Game Object** to appear in the *Game World*.

The second problem is that when a new player joins the *Game World* late, with a new **Peer**, all of the **Events** that were received by the **Peers** since the game began has to be replayed for the new **Peer**.

These two problems are not unique to the **Event-Database Architecture**. Many commercial *game-engines* face the same problems because these use a **Peer to Peer Network Architecture**. Even though the authors of the *game-engines* use the terminology of a **Client Server Network Architecture** which makes the matter confusing.

One common solution to the first problem is to ensure that the effect of **Events** on the **Game Database** is not random but deterministic i.e. predictable. By, for example, making sure that all the **Peers** start with the same value or **RANDOM SEED** when generating random numbers. Random numbers are generated in computers using a value and a mathematical formula to generate pseudo random numbers. When you pass the value into the formula the result is a pseudo random number. And when you pass this pseudo random number back into the formula you get another pseudo random number and so on and so on. So long as you start generating pseudo random numbers from the same value, or **Random Seed**, the sequence of random numbers from the formula will always be the same i.e. predictable.

Another common solution to the second problem is for the **Events Host** to replay all of the **Events** in the **Events History Record** on a new **Peer**, that connects to the **Central Host**. To get the *Game World* on that new **Peer** up-to-date and synchronised with the rest of the *Game Worlds* on the network. But this can take a long time if the queue of **Events** is very long.

Another solution to the second problem is for the **Central Host** to just synchronise the **Game Database** on the new **Peer** that connects to it. By sending all of the changes to its local copy of the **Game Database**, to the **Database Host** of the new **Peer**. This too can take a long time if there have been a lot of changes.

Another solution to the second problem is for the **Central Host** to synchronise the **Game Database** on the new **Peer** by only sending a select minimum number of changed *Records* or *Fields* to the new **Database Host**. The *Database Records* or *Fields* selected would be chosen by some rule set by the *Game Programmers* and *Database Administrators.* To only include the minimum number of important *Records* required to achieve synchronisation.

For example, you could have a rule that only the *Database Records* of **Game Objects** close to the new player should be synchronised in the new **Game Database**. All of the *Database Records* of **Game Objects** that were far away from the new player should be ignored. Or you could have a rule that only the *Database Records* or *Fields* that had a special tag or word associated them, would be replicated.

For example, you could have a **Database Meta Data Records** or **DATABASE TAG RECORDS** which tag other *Records* with words which give them special properties. Such as the ability to be replicated across a computer network. You would use these **Database Tag Records** to ensure that only **Game Objects** whose physical properties were being updated in the *Game World* by the **Physics Host** were replicated. But the *Database Record* of the **Camera Objects** being used to display the *Game World* were not replicated. Since the view of the *Game World* should be different for each player but the **Game Objects** in the *Game World* should not. There is an example of a *Database Table* that would list external *Database Records* (outside of that *Table*) that would be replicated in the **Game Database** across the network in Table 3.3. This includes examples of the external *Database Records* being referred to in the *Database Table.*

Another example you could have a rule that the **Database Tag Records** would list all the *Database Fields* in a *Record* that should be replicated. You would use this to ensure that only the position, and not the speed or acceleration, of **Game Objects** were replicated across a computer network. To ensure synchronisation of

---

TABLE 3.3

Example of How *Database Records* Could Be Tagged to Be or Not to Be Replicated across a Computer Network

List ID	Replicated
Replication Tag List	3D Physics List, Warrior 3D Player Object, Thief 3D Player Object, Forest Sector Object, Forest Tree Object 1, Forest Tree Object 2, Forest Tree Object 3, Small Bush Object 1, Small Bush Object 2, Large Boulder Object, Sky Object.

List ID	Non-Replicated
Non-Replication Tag List	3D Camera List, Side View Camera Object, Game Logo Camera Object.

List ID	3D Point Object IDs
3D Physics List	Warrior 3D Player Object, Thief 3D Player Object, Forest Sector Object, Forest Tree Object 1, Forest Tree Object 2, Forest Tree Object 3, Small Bush Object 1, Small Bush Object 2, Large Boulder Object, Sky Object.

List ID	3D Camera Object IDs
3D Cameras List	Side View Camera Object, Game Logo Camera Object.

those **Game Objects** in the *Game World*. There is an example of a *Database Table* that would list external *Database Fields* (outside of that *Table*) that would be replicated in the **Game Database** across the network in Table 3.4. This includes examples of the external *Database Fields* being referred to in the *Database Table*.

These tags would be manually placed by the *Game Programmers*, *Game Designers*, *Sound Designers or Engineers* or *Database Administrators*. Depending on the *data* they believed should be replicated across the network.

All these solutions depend on the *data* between transmitted between the computers or **Peers** in the **Peer-to-Peer Network** being reliable or compensatory. That is

TABLE 3.4

Example of How *Database Fields* Could Be Tagged to Be or Not to Be Replicated across a Computer Network

List ID	Replicated	Fields
Replication Tag List	Warrior 2D Player Object, Thief 2D Player Object., Mage 2D Player Object, Cleric 2D Player Object	X, Y, Angular Position

List ID	Non-Replicated	Fields
Non-Replication Tag List	Warrior 2D Player Object, Thief 2D Player Object., Mage 2D Player Object, Cleric 2D Player Object	X Speed, Y Speed, X Accel., Y Accel., Angular Speed, Angular Accel

Object ID	Game Object Code	Mass	X	Y	X Speed	Y Speed
Warrior 2D Player Object	Warrior 2D Player Code	100	320	240	0	0
Thief 2D Player Object	Thief 2D Player Code	90	90	51	0	0
Mage 2D Player Object	Mage 2D Player Code	80	121	128	0	0
Cleric 2D Player Object	Cleric 2D Player Code	50	192	200	0	0

Object ID	X Accel.	Y Accel.	Angular Position (Deg.)	Angular Speed (Deg./Sec.)
Warrior 2D Player Object	0	0	0	0
Thief 2D Player Object	0	0	0	0
Mage 2D Player Object	0	0	0	0
Cleric 2D Player Object	0	0	0	0

Object ID	Angular Accel. (Deg./Sec./Sec.)	Collision Boundary ID	Proximity Boundary ID	Collision Event ID
Warrior 2D Player Object	0	Warrior 2D Low Res. Polygon	Warrior 2D Low Res. Polygon	Warrior Collis. Event
Thief 2D Player Object	0	Thief 2D Low Res. Polygon	Thief 2D Low Res. Polygon	Thief Collis. Event
Mage 2D Player Object	0	Mage 2D Low Res. Polygon	Mage 2D Low Res. Polygon	Mage Collis. Event
Cleric 2D Player Object	0	Cleric 2D Low Res. Polygon	Cleric 2D Low Res. Polygon	Cleric Collis. Event

to say when an **Event**, or the properties of **Game Objects** or *Database Fields*, was replicated across the network, there was either no loss of *data* or packets containing the **Events** or properties or *Database Fields* to be replicated, from the **Central Host** to the **Events Hosts** or the **Database Hosts** on the network. And the **Central Host** receives confirmation for every packet it sends out. Or if there were a loss, then the **Central Host** automatically re-transmits the packets lost. Or the **Central Host** compensates for the loss, by transmitting later packets that contain the accumulation of all prior **Events** or changes to properties or *Database Fields*. And synchronise the *Game World* on the **Peers** to the latest state.

These solutions are used in most commercial *game-engines* in the Computer Games industry. But these do not have the advantage that the **Event-Database Architecture** has which is that it is based on a *Relational Database* and a *Relational Database Management System*[44] that the **Database Host** uses to control access to that *Database*. Some *Relational Database Management Systems* come in a **Distributed Form**. So they automatically allow multiple Users to synchronise copies of a *Relational Database* across a computer network. If you use one of these Systems to build your **Database Hosts**, then you would not need to come up with your own solution to the problem of synchronising the **Game Database** across the computer network. The *Relational Database Management System* will do this automatically for you.

Another advantage that the **Event-Database Architecture** has over these commercial *game-engines* is that it comes in a **Multi-User Distributed Form** based on a **Client Server Network Architecture**. As explained earlier, the commercial *game-engines* actually implement a **Peer-to-Peer Network Architecture** and use the terminology of a **Client Server Network Architecture** when describing the results. Even though they have not really implemented that **Network Architecture**. There have been attempts by third-party *Software Developers* to recreate these *game-engines* with a real **Client Server Network Architecture**. But these have not been successful or widely adopted.

NOTES

1. *Game Controller.* A device used to control the User Interface, including the player and other characters, of a game.
2. *Sound stream.* A recorded sample of sound encoded in a special data format.
3. *Game time.* The number of seconds since a game was started.
4. *Vertices.* The point at which two or more sides of a shape meet. Three vertices are used to form triangles, which make up a 3D model. Four vertices are used to form a quadrilateral, which marks the position of a rectangular 2D image.
5. *Polygon.* A closed plane shape, with three or more sides. Triangles are used to make up a 3D model. Quadrilaterals are used to mark the position of a rectangular 2D image.
6. *Vector.* The magnitude and direction of a physical quantity e.g. force, speed etc. *Normal Vector.* A Vector with a magnitude of 1 that simply specifies the direction in which a 2D or 3D surface is facing. See Glossary.
7. *Texture.* A 2D image which is used to fill in a polygon. Only the region of the image specified by the Texture coordinates, of the polygon, is used to fill it in.
8. *Texture coordinates.* A set of points describing the region of an image which should be used to fill in a polygon. There are the same numbers of points as there are vertices in the polygon. Each point corresponds to one, unique vertex.

9. *X Position.* The position of a body along the X axis in a 2D or 3D space.
10. *Y Position.* The position of a body along the Y axis in a 2D or 3D space.
11. *Z Position.* The position of a body along the Z axis in a 2D or 3D space.
12. *X Angular Position.* The rotation of a body, in a local 2D or 3D space with an origin at its centre of mass, around the X axis, in a plane perpendicular to the axis or the ZY plane.
13. *Y Angular Position.* The rotation of a body, in a local 2D or 3D space with an origin at its centre of mass, around the Y axis, in a plane perpendicular to the axis or the ZX plane.
14. *Z Angular Position.* The rotation of a body, in a local 2D or 3D space with an origin at its centre of mass, around the Z axis, in a plane perpendicular to the axis or the XY plane. In 2D space the Z axis does not exist and it's just an imaginary axis extending out from the 2D plane.
15. *Frame.* A single image in an animated sequence. A single image of an animated world.
16. *Software Rendering.* Rendering items in 2D or 3D space using a Central Processor and main memory in a computer system.
17. *Hardware Rendering.* Rendering items in 2D or 3D space using a specialised Graphics Processor and Graphics memory in a computer system.
18. *Near and Far focal length.* The closest and furthest distance of the visible area or volume in front of a camera.
19. *Field of View.* The angle between the left hand side and the right hand side of the visible area or volume in front of a camera.
20. *X Speed.* The speed of a body along the X axis in 2D or 3D space.
21. *Y Speed.* The speed of a body along the Y axis in 2D or 3D space.
22. *Z Speed.* The speed of a body along the Z axis in 3D space.
23. *X Acceleration.* The acceleration of a body along the X axis in 2D or 3D space.
24. *Y Acceleration.* The acceleration of a body along the Y axis in 2D or 3D space.
25. *Z Acceleration.* The acceleration of a body along the Z axis in 3D space.
26. *X Angular Speed.* The rotational speed of a body around the X axis in 3D space.
27. *Y Angular Speed.* The rotational speed of a body around the Y axis in 3D space.
28. *Z Angular Speed.* The rotational speed of a body around the Z axis in 2D or 3D space.
29. *X Angular Acceleration.* The rotational acceleration of a body around the X axis in 3D space.
30. *Y Angular Acceleration.* The rotational acceleration of a body around the Y axis in 3D space.
31. *Z Angular Acceleration.* The rotational acceleration of a body around the Z axis in 2D or 3D space.
32. *Sound channel.* A component of computer generated sound, which can play back a sound (given the sound envelope, i.e. the shape of the sound wave, or a sound stream) independently, or mixed with other sound channels.
33. *Analogue device.* A device which produces data that measures a continuously variable, physical quantity e.g. The rotation of a Joystick about its X, Y or Z axes, the pressure applied to a button.
34. *Digital device.* A device which produces data that measures a binary, physical quantity e.g. A Joystick being moved to the left or right, a button being pressed or released.
35. *Operating System.* A software that controls how other software share resources on the same computer hardware.
36. *Software Application.* A software program that is used directly by a Software User, through a User Interface, to solve a problem.
37. *Process (in an Operating System).* A software program or routine that is running in its own space in computer memory. When it is part of an Operating System or Software Application, the System or Application can temporarily interrupt it, to allow other Processes to share the resources on the computer hardware. Before the Process is resumed.
38. *Thread (in an Operating System).* A sub Process generated from another Operating System Process which shares the same space in computer memory as its parent. This simplifies and speeds up the communication between the two Processes.

39. ***TCP/IP Address.*** Transmission Control Protocol or Internet Protocol Address is a unique word, normally made up of 4 numbers separated by dots, used to identify the source or destination of a message, being sent between two computers on a network.

40. ***Port Number.*** A number that represents a channel through which messages can be sent or received by a computer on a network. Several messages may be sent or received in parallel on the different channels on the same computer.

41. ***Username.*** The unique name of a Software User used to identify that User and the resources e.g. files, Threads or Processes that they own in an Operating System or Software Application.

42. ***Password.*** The unique word that only a Software User knows and uses to authenticate their access to resources available on an Operating System or Software Application, that they own.

43. ***Authentication Token.*** A unique encrypted word that is generated by a computer, from a Username and Password, to authenticate that User's access to resources available on an Operating System or Software Application, that they own.

44. ***Relational Database Management System.*** Software that create, edit and query a Relational Database. It normally includes a standard programming language, Structured Query Language or SQL, that allows you to query the database.

4 The Software Production Process

In the **Event-Database Architecture**, the flow of a game would be controlled by a series of **Events**. This would give the *Game Producers* and *Designers* the ability to easily extend or modify an incomplete *game design* to make minor changes. Using **Events** would produce a lot more *game modules* to manage than usual. This would be because any *software procedure*, which would have had *logic branches* in it, would now possibly be spread across two *modules*, with an **Event** linking the two. All of these *modules*, although small and simple, would have different properties and uses. All of this information would have to be managed.

As already mentioned, a *Database Administrator* would play a major role in managing this information. Each **Event** and *game module* (or **Game Object**) would have a *Database Record*, with its properties. From these *Records*, the *Administrator* would know something about the use of each. Any documentation the *Administrator* kept, about the *Database*, would be useful to people trying to modify the game.

The *Game Programmers* too would have a large understanding of the many **Game Objects** available, since they would write each *module*. They would be released from having to make minor changes to the game, when the *game design* changed. They could use this opportunity to spend more time planning, documenting and implementing major changes, using the network of **Events** and **Game Objects** instead.

But no one *Programmer*, or *Database Administrator*, would know all that could be possible with the set of **Events** and **Game Objects**. They would soon lose track, especially when this set started to change and get large. Nor would all the changes in a *game design* be met by any given set of **Events** and **Objects**.

The solution to these problems would be a process for producing games, which systematically dealt with changes to a *game design*. Beginning with the initial design, followed by intermediate changes and implementations, and ending with the final game, the process would ensure that the same simple method was used at each stage.

One example, of such a process, could begin with the assumption that no **Host modules** have been built. Instead, the **Host modules**, the **Game Objects**, the **Events** and the **Game Database** will all be built as part of the process. The process would follow these steps:

1. A meeting with the staff would be organised where you would announce the game and the decision to use the **Event-Database Architecture** and the **Event-Database Production Process**. You would give an overall vision for the production process. You would get the staff to collectively agree on the maximum time to investigate a task before implementing that task during the process.

DOI: 10.1201/9781003502784-4

2. The *Database Administrators, Game Programmers, Game Artists, Sound Designers, Game Designers and Game Testers* would produce a feasibility study. That would implement a minimal game based on the **Event-Database Architecture** on the computer hardware the game was targeted for. This may or may not include a narrow but deep cross section of the *Game World*, to demonstrate its feasibility also called a 'Vertical Slice'. The steps of the feasibility study to build a cross section of the game are almost exactly the same as the steps to build the whole game. The only difference being that it only requires steps (3) to (35).

3. Investigate how long it would take to write the *game design, technical design, data design* and *tools design* and give an estimate of how long it would take to complete these designs based on the investigation.

4. Record when the creation of the *game design, technical design, data design* and *tools design* started in the *game design*. This should be the first entry in the *game design*.

5. The *Game Designers*, with help from the *Game Artists* and *Sound Designers*, would complete the rest of the *game design* of the whole game or the cross section of the game, in the case of a feasibility study.

6. The *Game Programmers* would produce a *technical design*. This document would describe the **Host Modules, Game Objects** and the **Events** required to build the first version of the *game design*, the whole game or a cross section of the game in the case of a feasibility study. It would include the techniques that would be used to implement the features in the *game design*. This would include either an explanation of the principles of the **Event-Database Architecture** or a reference to another document which included this explanation. But the *technical design* would not include the *data* or the tools that were to be used.

7. The *Database Administrator* would produce a *data design*.[1] This would be written after consultation with the *Game Programmers, Game Artists, Game Designers, Sound Designers* and *Game Testers*. This would describe all the *Records* that would be contained in the **Game Database**, and the *data* contained in these *Records*. These would also include any *data* the **Host Modules, Game Objects** or **Events** required. It would include a description of the *data* produced by the software tools that were to be used. But it would not include a description of the tools.

8. The *Game Programmers*, with help from the *Game Artists, Sound Designers, Game Designers, Game Testers* and *Database Administrator*, would produce a *tools design*.[2] This would describe all the third-party and custom tools, including the RDBMS, that they would require to create and manage all the data and Game Database described in the *data design*. It would also include the tools that would be used to write and build the game and the custom tools. And it would include any software repositories to archive the computer files used to build each version of the game or custom tools.

9. Record when the creation of the *game design, technical design, data design* and *tools design* ended.

10. Investigate the steps required to build or purchase the third party or custom tools described in the *tools design* and give an estimate of how long it would take to based on the investigation.
11. Record the time when the building started in the *game design*.
12. The Game Producers, Game Programmers and Database Administrators would build or purchase the third party or custom tools described in the *tools design*.
13. Record the time when the building ended.
14. Investigate how long it would take to get or build the first set of *data* required by the **Game Database** and give an estimate of how long it would take based on the investigation.
15. Record when the building began in the *game design*.
16. The *Game Designers*, *Game Artists* and *Sound Designers* would get or build the first set of *data* using the third-party or custom tools.
17. Record when the building ended in the *game design*.
18. The *Database Administrator* would investigate the steps required to build the **Game Database** with the RDBMS from the first set of *data* and give an estimate of how long it would take.
19. Record the time when the building began in the *game design*.
20. The *Database Administrator* would create the **Game Database**. From the first set of *data* created in the preceding steps.
21. Record the time when the building ended in the *game design*
22. Investigate the steps it would take to build the **Events Host** and give an estimate of how long it would take based on the investigation.
23. Record the time when the building of the **Events Host** started in the *game design*.
24. Build the **Events Host**.
25. Record the time the building ended in the *game design*
26. Repeat (2), (3) and (4) for the **Database Host, Objects Host, Physics Host, Graphics Host, Sounds Host, Game Controllers Host** and **Central Host**.
27. Investigate the steps it would take to build first set of **Game Objects** which were required by the **Game Database**. And give an estimate of how long it would take based on the investigation.
28. Record the time when the building of the **Game Objects** started in the *game design*.
29. Build the **Game Objects**.
30. Record the time the building ended in the *game design*.
31. Investigate the steps it would take to assemble the **Host Modules, Game Database** and **Game Objects** to build the first version of the game and give an estimate of how long it would take based on the investigation.
32. Record when the assembly began in the *game design*.
33. Assemble the game.
34. Record when the assembly ended in the *game design*.
35. The *Game Testers* would test the **Events** of the first version of the game. This test would be carried out against the initial *data design*. And it would simply check that all the **Events** produce the expected results

in the *data design*. If a *Game Tester* cannot tell the expected results of an **Event** from its description in the *data design*, then either the **Event** should not exist. Or its name needs to be changed. Or its description needs to be changed. Otherwise the language on the project will degenerate.

36. If a change to the *game design* were required, the *Game Designers* should investigate with the *Database Administrator* and *Game Programmers* whether it would be feasible to implement the change, using the current set of **Events**, **Game Objects**, *Database Tables*, *Database Records* and *Database Fields*.

37. If it were possible, the *Game Designers* would give an estimate of how long it would take to make that change, based on the investigation.

38. Record the time when that change began in the *game design*.

39. Implement the change by editing the *Database*, with one of the custom tools described in the *tools design*.

40. Record the time when that change was completed in the *game design*.

41. If the change were not possible, then the *Game Designers* would investigate the steps required to implement the closest solution possible, using the existing items in the **Game Database**, and give an estimate of how long this would take.

42. Record the time when that change began in the *game design*.

43. The *Game Designers* would implement the closest solution possible, using the existing items in the **Game Database**.

44. Record the time when that change ended in the *game design*.

45. The *Game Designers*, *Game Artists*, *Sound Designers* and *Game Programmers* would investigate what new **Events**, **Game Objects** and *data* were required to add this new feature and give an estimate of how long these would take to build. This would include any new additions required, from the **Host Modules**, which followed the principles of the **Architecture**.

46. Record the time when these new additions for this new feature began in the *game design*.

47. The *Game Artists*, *Sound Designers* and *Game Designers* would design each new artwork, sound or other *data*. The *Game Programmers* would write a design of each new **Game Object**. Or they would modify the design of each affected **Host Module** to include any new additions required. These new designs, or additions, need only to include a description of any new set of **Events**, *Database Tables*, *Records* and *Fields* required from the **Game Database**.

48. The *Game Artists*, *Sound Designers* and *Game Designers* would build their new *data*.

49. The *Database Administrator* would verify that this new *data* did not already exist in the **Game Database** and were well-defined. Before adding their definition to the *data design*, and adding that *data* for the new **Events** and **Game Objects** to the **Game Database**. But these would all initially be inactive. That is, no existing **Game Objects** or **Host Modules** would use this new *data*.

50. The *Game Programmers* would write and build the new **Game Objects**. And these would be added to the set used by the **Objects Host**, to respond to the new **Events**. Or they would rewrite and rebuild the extended **Host Modules**, if that were required.

51. The *Game Testers* would test each new **Event**, **Game Object** and *data* separately, using one of the custom tools described in the *tools design*. Or they would test the additions made to each **Host Module** separately. These tests would be carried out at least against the *data design*. This would require a custom tool to be included in the tools design, which allowed the *Game Testers* to either send any **Primary** or **Secondary Event** they chose, at any time during the game. Or it would allow them to improvise with any **Object** they chose, at any time. Or it would allow them to modify any *Field* they chose, in the **Game Database**, at any time.

52. The *Game Designers* would implement the new feature required, using the new set of **Events**, **Objects**, *data* or additions to the **Host Modules**. And this would be done by editing the **Game Database**.

53. Finally, *Game Testers* would test the new feature.

54. Record the time when the new additions for this new feature were completed in the *game design*.

55. At the end of production or when a milestone was reached, after it had been decided that there would be no more changes to the *game design*, the *Game Testers* would go through the *Database*. And they would gather a list of the **Primary Events** used.

56. The *Game Testers* would produce a list of features of the game, based on the description of these **Primary Events** in the *data design*.

57. The *Game Testers* would test those features against the final game. And it would simply check that all the **Events** produce the expected results in the *data design*. If a *Game Tester* cannot tell the expected results of an **Event** from its description in the *data design*, then either the **Event** should not exist. Or its name needs to be changed. Or its description needs to be changed. Otherwise the language on the project will degenerate. If the **Events** did not produce the expected results, then the game would be edited using the previous steps.

Compare this production process with the *Software Evolution Process* used to build Computer Games. In the worst case, this *Software Evolution Process* would begin with no discernible phases at all. That is, there would be no phase involving the production of any designs. Be it a *game design*, *technical design* or other documentation, these would simply be bypassed. Instead, the production of the game would proceed in an ad hoc fashion, virtually instantaneously, with many overlapping phases of the *software production life cycle*. The production of *software modules* by the *Game Programmers*, artwork by the *Game Artists*, sounds by the *Sound Designers* and other *data* by other staff would all begin almost straight away.

There would be only an informal meeting between the staff where they would be shown either some sketches of a plan; describing the main tools that were to be used, the general theme of the game, the major lessons learnt from the last project, which the *Software Developer* wanted to avoid. Or they would be shown the demonstrations of a competing product, which the *Software Developer* wanted to emulate. Or they would be shown a previous game made by the *Software Developer* which they were going to improve upon. And after a brief verbal discussion about the initial expectations from each of the staff, the production would begin. Indeed, some of the staff may be left with no idea at all what their role was going to be. And if they enquired, they would merely be given verbal reassurances that it would all become clear at a later date.

In the best case, the *Software Evolution Process* would initially begin with the same phases as the *classic software production life cycle*. That is, it would begin with the first three phases. There would be some form of an analysis of the requirements of the software, along with a feasibility study, a *game design* and a *technical design*. But, after that, the process would descend, in earnest, into the same ad hoc process as the worst case.

Therefore, there would be no phase like step (7), where all the *data* currently in the game, and used by the tools, would be documented. Instead, the *data* would be introduced, especially new ones with increasing frequency towards the end of production, without any ceremony or documentation.

There would be no phase like step (8), where all the tools that were currently being used would be documented. Some of the custom tools would be referred to, briefly, in the *technical design*. But, in principle, custom tools would be introduced into the process on demand; while the game was being built, without documentation.

Of course, there would be no phases in the *Software Evolution Process* involving a *Database Administrator*. Instead, various members of the staff would be responsible for the *data* which they produced. They would each independently be responsible for merging their *data*, with the rest of the *Game data*. And this would often result in some of the *data* being incompatible with others, or replicated.

Furthermore, the *Game Designers* would have to interrogate these various members of staff if they wanted to change the *game design*. They would have to navigate through a maze of staff, to find out whether there already existed some *Game data* which they could modify to add a new feature. Each of these staff would only have a partial understanding of the *Game data* as a whole and would pass on each enquiry to someone else. This trail would inevitably lead to the *Game Programmers*. And they would, in turn, navigate through a maze of *software modules*, to see whether any *Game data* or *Abstract data* could be deployed for the task. Since, as has already been mentioned, these *modules* would be the only remotely comprehensible documentation kept up-to-date.

Despite the fact that these *modules*, like the *Game data*, would lack documentation, these computer files would nonetheless be the the de facto documentation for all the *data*. These would be the only items in the *Software Evolution Process* which place the *Game data* in context. And only after conducting such extensive investigations, to uncover this context, could the *Game Designers* be confident whether certain propositions were possible by editing the *Game data*. However, most of the time, they find this impractical. And they default on such investigations. Instead, they would prefer to either introduce new *data*, at the risk of duplicating any existing *data*. Or they would extend the first *data* which they encountered that remotely resembled whatever they were looking for. This would be done at the risk of making that *data* more complex, and consequently making all parts of the game dependent on that *data* more complex, and because of the greater complexity, more prone to errors..

Not only the *Game Designers* would do this, but all the other staff who wanted to modify the game would take similar risks. Whereas, with the **Event-Database Architecture**, there would be only one table; one reference. This would be namely the **Game Database.** And they could simply refer to this, rather than navigate through the maze of items in the *Software Evolution Process*. They could simply go through a single, linear path of enquiry, through the *Database*, from one *Record* to the next. They would not have to navigate multiple paths of enquiry simultaneously; the majority of which lead to dead ends, while others merely go around in circles.

There would be no phase in the *Software Evolution Process* involving the *Game Testers*, prior to the end of production. The lack of documentation during the process would mean that there would be no designs, with which they could test any features, introduced into the game, against.

There would be no phase where the *Game Designers* could change any part of the flow of the game, all the components in the game and all the properties of these components, by editing a *Database*. There would be no phase where the *Game Designers* would conduct each and every change to the *game design*. The *game-editors* would allow the *Game Designers* to edit some components or *software modules* of the game. But the ad hoc nature in which features were introduced into the game, especially new *Game data*, would limit this capacity. Many components or *software modules* would be added to the game which could not be edited with the *game-editor.* Since those who introduced these *software modules* would either feel it unwarranted, neglect to demand or have their request that the *game-editors be* changed to allow you to edit these *software modules* refused. The frequency of such introductions would increase towards the end of the production, as more and more software modules are rushed through bypassing many steps of the *software production life cycle* and without regard whether you can edit these modules with a game-editor. Although the *Developers* of many commercial *game-editors* market these on the ability of the editors to edit any *game design*, you can only edit some these words are not necessary of the components of a *game-design.*

The steady decline in the effectiveness of the *game-editors*, as more *software modules* were introduced which could not be edited with the *game-editors*, would also mean that, towards the end, there would be no assessment of the feasibility of making changes to the *game design* with the *game-editors*. Instead, changes would be introduced without any assessment of its feasibility and effect on the software. These changes would be written directly into the game, by the *Game Programmers*. And there would be no attempt to assess whether these changes could have been done with the *game-editors* first.

There would be some phases of the *Software Evolution Process* where designs written or drawn up by the *Game Programmers*, *Game Artists*, *Sound Designers* and *Game Designers* would be documented. But, later on, as more and more changes were made to the *game design*, these phases would disappear.

However, there would always be phases, in the *Software Evolution Process*, where the *Game Programmers* would write *game modules* and build these. But, unlike in the **Event-Database Production Process**, all the changes to the *game* *modules*, *Game data* or *Abstract data* would be incorporated into one step. So that these could all be tested at once, simultaneously. Whereas, in the **Event-Database Production Process**, all these changes would be disabled by default. So that each one could be tested independently. This would help prevent errors in one of the new **Events**, **Actions**, **Game Objects**, *Database Tables*, *Records* or *Fields* becoming confused with errors in old items in the *Database*, or with each other.

Finally, the *Software Evolution Process* would have no phases where the *Game Testers* could test features of the game against a design. At best, the *Game Testers* would be instructed on what features to test by the staff who added them, and what to look out for. Even though the staff in turn would have nothing more to back up these instructions than their own vague memories of why they added these features. And at worst, the *Game Testers* would be left to their own intuition to decide what features to test, and what looks good and what does not.

To summarise, the **Event-Database Production Process** has four advantages over a normal *Software Evolution Process*.

Firstly, when there is a new change to the requirements of the *game design*, the *Software Evolution Process* produces a new set of *game modules*, *Game data* and *Abstract data* to meet those requirements. The **Event-Database Production Process** produces a new set of **Primary Events**, **Secondary Events**, **Game Objects**, *Database Tables*, *Database Records* and *Database Fields*. The second set has a greater tolerance to the changes in the *game design* than the first. By virtue of its members.

Secondly, during the production process and especially at the end, you can identify and test every member of the second set, every **Primary Event**, **Secondary Event**, **Game Object**, *Database Table*, *Database Record* and *Database Field*. And thus you can have greater *Quality Control* than in a *Software Evolution Process*. This can all be done from one source and one tool: the *Game Database*. But you cannot identify and test every member of the first set, every *game module*, *Game data* or *Abstract data*. In a *Software Evolution Process*, there is no such single source or tool.

Thirdly, there is a book that explicitly explains the **Event-Database Production Process** and the **Event-Database Architecture**. This is called

Event-Database Architecture for Computer Games: Volume 1, Software Architecture and the Software Production Process.

This gives you an understanding of the *software architecture* at the beginning and the end of the process. There is no book which explains the *Software Evolution Process* for Computer Games and can predict the *software architecture* it will produce. You can only see this at the end of the process.

Fourthly, the **Event-Database Production Process** and the **Event-Database Architecture** provide you with a *Relational Database* and a *Relational Database Management System* for building Computer Games. To manage and query the huge amounts of data that it takes to build and run modern Computer Games. The *Software Evolution Process* does not. In a production process involving, for example, about 300 staff, this data can be over 400 gigabytes in size and involve 3.5 million files. The *Software Evolution Process* breaks down on such a scale.

4.1 STEP 1: FEASIBILITY STUDY/VERTICAL SLICE

The **Event-Database Architecture** is just a mechanism for producing *software designs* and a software production process. As such it would not be dependent on any set of tools. There are many sets of tools that could be used to build the *software designs* and software production process based on it.

Before using the **Architecture** to produce any large *software design*, however, you could build a simple design that allowed you to test the basic features of the **Architecture**. This would help demonstrate the feasibility, or otherwise, of using that **Architecture** on a given computer hardware. This is one of the first steps

step (2)

of the **Event-Database Production Process** as described in the preceding subchapter, which is a feasibility study. The object of the feasibility study is to examine both the **Event-Database Architecture** and the **Event-Database Production Process**. The study will involve executing the most important steps of the **Event-Database Production Process**

steps (1)–(35).

The steps begin by designing a small but deep cross section of the *Game World*, known as a 'Vertical Slice', that represents a fraction, for example 10%, of the whole game

step (5).

And the steps end with a test of that 10% based on the **Event-Database Architecture**

step (35)

and a prognosis would be made about how long it would take to build the whole game.

To make this prognosis, times are gathered in the *game design*, of when each subtask begins and ends during the **Event-Database Production Process**. These times are gathered in these steps

4, 9, 11, 13, 15, 17, 19, 21, 23, 25, 28, 30, 32, 34.

And at the end, these times are used to get an overall time for how long it took to build 10% of the whole game. And that time, in turn, is used to make a prognosis of how long it would take to build the whole game. And that in turn is used to assess whether it is feasible to build the game, given the deadline for the project, using the **Event-Database Architecture**.

Now the smaller test of the 10% of the game is different from the bigger test of the whole game described in the steps of the preceding subchapter. The exact nature of this test is described in the next chapter.

4.1.1   Designing the Test

A test of the **Event-Database Architecture** would require three sets of items. It would require a set of computer software components to be built and assembled. It would require a set of computer hardware components, which would use the assembled software to perform the test, and allow you to interact with the game. And it would require a set of small steps that you would follow to test the different components of the **Architecture**.

The set of software components would include the **Host Modules** of the **Event-Database Architecture**. It would also include the additional *software modules* (i.e. **Game Objects**) that would allow you to see items in the game and interact with these. And, finally, it would include a **Game Database** that stored all the shared *Game data*, and *Abstract data*, the *software modules* would use.

You would need to build all eight **Host Modules** for the test. The **Events Host** would control the flow of the game. The **Database Host** would store and retrieve all the *data* that would be used. The **Objects Host** would use the **Game Objects** to respond to **Events**. The **Physics Host** would move the **Game Objects** around. It would include gravitational and frictional forces, in its software model. So that all **Objects** fell down onto the ground and slowly came to rest whenever these were moved. The **Graphics Host** would display these **Game Objects**. The **Sounds Host** would play any accompanying audio for the test. The **Game Controllers Host** would read the position of the *Game Controllers* and allow you to use these to interact with the game. And, finally, the **Central Host** would be used to control and synchronise all the other **Host Modules**.

Along with these *modules*, you would need to build eight additional *software modules* in all. Three of these would be 2D **Game Objects**, which would be used to display 2D images, and demonstrate the movement of 2D items. Another three would be 3D **Game Objects**, which would be used to display 3D models, and demonstrate the movement of 3D items. One 2D **Camera Object** would be required to display the 2D items. And, finally, one 3D **Camera Object** would be required to display the 3D items.

The movement of the first 2D **Game Object** would be guided by a *Game Controller*. The **Game Object** could be any solid, arbitrary shape. But its colour would initially be green. And it would have a circle forming a boundary around it that would be used to detect its collision with other **Objects**. It would change its colour to red each time this happened. It would appear on the screen when the *digital devices*, of the *Game Controller*, were connected to the computer. And it would disappear when the same *devices* were disconnected. It would move up or down, left or right, when the *analogue devices*, of the *Game Controller*, were moved. This would either be when one *analogue device*, with horizontal and vertical axes, was moved. Or the **Game Object** would move, in both sets of directions, when two separate *analogue devices* (e.g. a button and some other *device* with one axis) were moved. It would also stop moving when these same *analogue devices* were stopped.

This 2D **Game Object** would respond differently when the *digital devices*, on the *Game Controller*, were pressed. These *devices* would be five different buttons. When the first button was pressed, the speed of the **Game Object** would be subsequently affected by the movement of the *analogue devices*. By default, these movements would only affect its position. When the second button was pressed, the movement of the *analogue devices* would revert back to affecting its position. When the third button was pressed, the scale of this effect on its speed or position would be increased. When the fourth button was pressed, the scale of the effect would be decreased. And, finally, when the fifth button was pressed, the game (i.e. the software being tested) would shut down, after an error had been briefly displayed. This error would simply state that the game was about to shut down, confirming the button was pressed.

The second 2D **Game Object** required for the test would not be directly controlled. It would only move when the controlled 2D **Game Object** collided with it. It also could be any solid, arbitrary shape. But its colour would be yellow. The **Object** would have a trapezium acting as an invisible boundary around it that would be used to detect its collisions. After each collision, with any **Object**, it would change the colour of the controlled **Game Object** to blue. And it would have twice the mass of that **Object**.

This **Game Object** would play three different sounds. When another **Object** came within close proximity of it, it would play one unique sound. It would play a second sound when the **Object** collided with it. And it would play a third sound, when the second sound had finished.

The third 2D **Game Object** would be used to limit the movement of the other two **Objects**, during the test. It would have the shape of a hollow rectangle. It would be predominantly transparent, except for the white outlines which would be clearly visible along its edges. The boundary around the **Object** would be a rectangle, which would deflect the other two **Objects** back inwards, from the outer edges. It would be large enough to contain both **Objects** and free space, within which these could move. The **Game Object** would have a mass so large that it would not move when the other two collided into it.

These two **Game Objects** would be placed next to each other within this boundary, when the test began. The **Objects** would be close enough to each other so that both would be seen. But, at the same time, there would be enough distance between the two so that the **Objects** were not in close proximity to each other. When one of these **Objects** hit the boundary, the third **Game Object** would flood all the *sound channels* with a unique sound. This sound would have a higher priority than all other sounds used during the test.

As with the first 2D **Game Object**, the movement of the first 3D **Game Object** would be guided by a *Game Controller*. But this second *Game Controller* would be different from the one used to control the 2D **Game Objects**. The **Game Object** would have a sphere acting as a boundary around it that would be used to detect its collision with other **Objects**. The **Game Object** would appear when the *digital devices*, of the *Game Controller*, were connected to the computer hardware. And it would disappear when the same *devices* were disconnected.

Similar to the first 2D **Game Object**, the **Object** would move backwards or forwards, left or right along the floor, when the *analogue devices*, of the *Game Controller*, were moved. This would either be when one *analogue device*, with horizontal and vertical axes, was moved. Or the **Game Object** would move, in both sets of directions, when two separate *analogue devices* (e.g. a button and some other *device* with one axis) were moved. It would also stop moving when these same *analogue devices* were stopped.

Again, like the first 2D **Game Object**, the **Object** would respond to five different digital buttons, on the *Game Controller*, when these were pressed. When the first button was pressed, it would switch the effect of the movement, of the *analogue devices*, to the speed of the **Object**. By default, these movements would only affect its position. When the second button was pressed, the movement of the *analogue devices* would switch back to affecting its position. When the third button was pressed, the scale of this effect on its speed or position would be increased. When the fourth button was pressed, the scale of the effect would be decreased. And, finally, when the fifth button was pressed the game would be shut down.

The second of the 3D **Game Objects** would not be directly controlled, during the test. It would only move when the controlled 3D **Game Object** collided with it. The **Game Object** would have a cube forming a boundary around it that would be used to detect these collisions. The cube would have two ramps inclined against its right and left vertical sides. These ramps would also be part of the boundary around it. And the **Game Object** would have twice the mass of the controlled 3D **Game Object**.

The **Game Object** would play three sounds, just like the uncontrolled 2D **Game Object**. When another **Object** came within close proximity of it, the **Game Object** would play a sound. This would be the same sound you would hear when two 2D **Game Objects** came near each other. Similarly, when it collided with another **Object**, the **Game Object** would play another sound. This would be same sound you would hear when two 2D **Game Objects** collided. And when this sound had ended, it would play the same sound you would hear after the collision of 2D **Game Objects**.

The third 3D **Game Object** would be used to limit the movement of the other two 3D **Game Objects**, during the test. It would have the shape of a hollow cube. It would be predominantly transparent, but with white lines clearly visible along its edges. It would also have an opaque square floor, made up of an alternating sequence of black and grey parallel lines. When the **Object** was placed in the 3D world space, the lines along this floor would be parallel to the Z-axis. The boundary around the **Game Object** would be a cube that would deflect any other **Object** back inwards, when these collided into one of its sides. It would be large enough to contain the other two **Game Objects** and free space, within which these could move. It would also have a mass so large that it would not move, when the smaller **Game Objects** collided into it.

The smaller **Game Objects** would be placed within this boundary, when the test began. As with the 2D **Game Objects**, both **Objects** would be close enough to each other to be visible. But, at the same time, there would be enough distance between both that the **Objects** were not in close proximity to each other. When an **Object** hit the boundary, the third 3D **Game Object** would flood all the *sound channels* with the same sound. This would be the same sound you would hear when one of the 2D **Game Objects** hit the boundary that confines those **Objects**.

So that you may be able to see the two confined 2D **Objects**, when the test began, the 2D **Camera Object** would be carefully positioned. Its initial position and area of visibility would be such that you would see both **Game Objects**. But, at the same time, it would not be possible for the camera to see the entire space within which these **Objects** could move. This would enable you to clearly see the movement of the camera across this space. After each collision, between these **Game Objects**, the camera would reposition itself above the uncontrolled **Game Object**, looking directly at it.

Similarly, the 3D **Camera Object** would be carefully positioned when the test began. Its initial position and area of visibility would be such that you would see the two confined 3D **Game Objects**. But, at the same time, it would not be possible for the camera to see the entire space within which these **Objects** could move. After each collision, between these **Game Objects**, the camera would reposition itself above and in front of the uncontrolled 3D **Game Object**, looking directly at it.

The information that the **Camera Object**, and all other *software modules*, would use to perform the test would be held in a **Game Database**. In total, only 86 *Records* would be required in this *Database*, for all the *modules*.

The **Events Host** would require 27 *Records*. One *Record* would be required for each of the 12 standard **Primary Events** of the **Architecture**. One would be required for each **Proximity Event** of the three 2D **Game Objects**. One would be required for each **Collision Event** of the three 2D **Game Objects**. Likewise, six would be required for the **Proximity Events** and **Collision Events** of the 3D **Game Objects**. One *Record* would also be required to hold the list of delayed **Events**. And another **Priority Events List Record** would be required to decide between two conflicting **Secondary Events**, based on the priority of each one. For example, the two **Secondary Collision Events**, of the two confined 2D **Game Objects**, would both need to be assigned equal priorities. So that sometimes the colour of the controlled 2D **Object** would turn red after a collision between the two. And sometimes, it would turn blue after each collision. And one **Events History Record** would be required to hold a history of all **Events**.

The **Database Host** would require four *Records*, in order to test its debugging features. One **Database Log Record** would be required to hold the list of *data* that would be monitored. Although several pieces of *data* could be monitored, for this simple test however, there would only be one entry in the list. And that would be the Primary Key of the *Record* which held the properties of the uncontrolled 3D **Game Object**. Another *Record* would be required to keep a log of the changes to this *data*. Another *Record* would keep a list of *Records* loaded into the computer memory. Another *Record* would keep a list of *Records* unloaded from memory.

The **Objects Host** would require three *Records*. One would hold the list of **Game Objects** that would be loaded into, and used, in the computer memory. One would hold the list of **Game Objects** that had a critical error or Crashed when responding to an **Event**. One would hold the list of times that these critical errors or Crashes occurred.

The **Physics Host** would only require three *Records*. One *Record* would be used to hold a selected list of 2D **Game Objects**. And the other would hold a selected list of 3D **Game Objects**. These **Objects** would be those whose position, speed and acceleration would be updated during each *Unit of game time*. Another *Record* would hold the properties of the **Master Physics Object** which are the parameters of the physics including the strength of the force of gravity, and resistance in solids, liquids and gas.

The **Graphics Host** would require 24 *Records*. Two *Records* would be required to hold two 2D polygons. Another two *Records* would be required to hold two 3D models. One *Record* would be required to hold a *Texture*. Another six *Records* would be required to hold the six *Texture coordinates* for the six **Game Objects**. One *Record* would be required to hold the list of 2D **Game Objects** that would be displayed. Another would be required to hold the list of 3D **Game Objects** that would be displayed. Another six *Records* would be required to hold the projection of the six **Game Objects** through the Cameras. Another two *Records* would be required to hold the list of 2D projections and the list of 3D projections. And another two *Records* would be required to hold the 2D **Camera Object** and the **3D Camera Object**. And another *Record* would be required to hold the list of Camera Objects.

The **Sounds Host** would require four *Records*. One *Record* would be used to hold the list of sounds that were waiting to be played. One *Record* would be used to hold the lists that were being played. One *Record* would hold the properties of the **Sound Microphone Object**. And one *Record* would hold the properties of the **Master Sound Speaker Object** through which all **Secondary Speaker Secondary Events** would pass through.

The **Game Controllers Host** would require six *Records*. Two **Device Group Records** would be required to hold the **Game Objects** which each group of *analogue devices*, on the two *Game Controllers*, would affect. Another two **Device Group Records** would be required to hold the **Game Objects** which each group of *digital devices*, on the two *Game Controllers*, would affect. One *Record* would be required to hold the properties of the **Master Player Object** that all **Events** related to players would pass through. And 14 **Device Sequence Primary Events Records** would be required to map the sequence of *analogue devices* or *digital devices* to **Secondary Events** that would be sent through the **Master Player Object**.

The **Central Host** would require only two *Records*. One would be required to hold the error displayed when the game was shut down. Another would hold the *Unit of game time* that would be used to operate the game. This would also hold when the game started and how much time had elapsed since it began.

The controlled 2D **Game Object** would require 12 *Records*. One *Record* would be required to hold its *Texture*. Another would be required to hold its *Texture coordinates*. Another would be required to hold the *vertices* of the quadrilateral that would display its image on the screen. And yet another would be required to hold the properties of the bounding shape that would be used to detect its collision with other **Objects**. One *Record* would be required to hold its unique properties, including its mass, position, speed and acceleration. One *Record* would be required to hold the projection of the **Game Object**, through the 2D camera. And six *Records* would be required to hold the properties of the **Secondary Events** it would receive. These would be sent when the first *Game Controller* was either connected, disconnected or its *devices* were moved, stopped or pressed. And another would be sent after the **Game Object** had collided with other **Objects**.

The uncontrolled 2D **Game Object** would require 13 *Records* in all. One *Record* would be required to hold its *Texture*. Another would be required to hold its *Texture coordinates*. Another would be required to hold the *vertices* of the quadrilateral that would display its image on the screen. Another would be required to hold the properties of the bounding shape that would be used to detect its collision with other **Objects**. And yet another would be required to hold the *vertices* and *Normal Vectors* of the bounding shape. One *Record* would be required to hold its unique properties, such as its mass, position and speed. One *Record* would be required to hold the projection of the **Game Object**, through the 2D camera, onto the screen. And three *Records* would be required to hold the properties of the **Secondary Events** it would receive. These would be sent when either another **Object** came within close proximity of it, collided with it or after the sound played during the collision had ended. A further three *Records* would be required, by the **Game Object**, to hold the sounds it would play when each of these **Events** occurred.

The hollow 2D **Game Object**, which would confine the other two **Objects**, would require five *Records*. One *Record* would be required to hold the properties of the bounding shape that would be used to detect its collision with the other **Objects**. Another would be required to hold the *vertices* and *Normal Vectors* of the bounding shape. One *Record* would be required to hold the unique properties of the **Game Object**, such as its mass and position. And one *Record* would be required to hold the only **Secondary Event** it would receive. This would be sent when another **Object** collided with it. A further *Record* would be required to hold the sound that it would play when this **Event** occurred.

The controlled 3D **Game Object** would require 12 *Records*, just like the controlled 2D **Game Object**. One *Record* would be required to hold its *Texture*. Another would be required to hold its *Texture coordinates*. Another would be required to hold the *vertices* of the 3D model that would display its image on the screen. And yet another would be required to hold the properties of the bounding shape that would be used to detect its collision with other **Objects**. One *Record* would be required to hold its unique properties, including its mass, position, speed and acceleration. One *Record* would be required to hold the projection of the **Game Object**, through the 3D camera, onto the screen. And six *Records* would be required to hold the properties of the **Secondary Events** it would receive. These would be sent when the second *Game Controller* was either connected, disconnected or when its *devices* were moved, stopped or pressed.

The uncontrolled 3D **Game Object** would require 13 *Records*, just like the uncontrolled 2D **Game Object**. One *Record* would be required to hold its *Texture*. Another would be required to hold its *Texture coordinates*. Another would be required to hold the *vertices* of the 3D model that would display its image. Another would be required to hold the properties of the bounding shape that would be used to detect its collision with other **Objects**. And yet another would be required to hold the *vertices* and *Normal Vectors* of the bounding shape. One *Record* would be required to hold the projection of the **Game Object**, through the 3D camera, onto the screen. And three *Records* would be required to hold the properties of the **Secondary Events** it would receive. These would be sent when either another **Object** came within close proximity of it, collided with it, or after the sound played during the collision had ended. It would play the same three sounds heard when these **Events** occurred for the uncontrolled 2D **Game Object**. So it would use the same three *Records*, which hold these sounds.

The hollow 3D **Game Object** would require five *Records*, just like the hollow 2D **Game Object**. One *Record* would be required to hold the properties of the bounding shape that would be used to detect its collision with other **Objects**. Another would be required to hold the *vertices* and *Normal Vectors* of the bounding shape. One *Record* would be required to hold the unique properties of the **Game Object**, including its mass, position and speed. And one *Record* would be required to hold the solitary **Secondary Event** it would receive. This would be sent when another **Object** collided with it. It would play the same sound heard when this **Event** occurred for the hollow 2D **Game Object**. So it would use the same *Record*, which holds that sound.

The 2D **Camera Object** would only require one *Record*. This *Record* would hold its properties, including its position and the size of its viewing area.

Finally, the 3D **Camera Object** would also require one *Record*. This *Record* would similarly hold its properties, including its position, orientation and the size of its viewing area.

The *Record* for the 3D camera would be the last of the *Records* that would be in the **Game Database**. If you were to add up all the *Records* required for each *software module*, you would end up with a total of 115 *Records*, for this test. But, six of these *Records* would be duplicates. These would namely be those *Records* used to hold the different sounds the 3D **Game Objects** would play. These would be the same sounds played by the 2D **Game Objects**. So the final total would be 109 *Records*.

The **Game Database** would be 17th and last software component that would be built for the test. There would be fewer hardware components required by comparison. There would only be four in all. The first of these would be a computer monitor that would display the game. The second would be a computer, which would be operated by the software, and be connected to the monitor. The computer would have a storage media that would be large enough to hold the *game software*, and the **Game Database**. The last of the hardware components would be two *Game Controllers* that would be connected to the computer.

Each *Game Controller* would be equipped with *analogue devices* and *digital devices*. The *analogue devices* would include either one *device*, with two axes. Or it would include two separate *devices*, which could be combined, and used as a substitute for a *device* with two axes. In the latter case, each of the *analogue devices* could either be a button, or some other *device* with one axis. The *digital devices* on the *Game Controller* would include at least five buttons.

Once the software components had been built and assembled, the results would be tested on the computer hardware. The test would follow these 49 steps:

1. You would disconnect all *Game Controllers* from the computer.
   [This would be in preparation for testing the ability, of the **Game Controllers Host**, to detect the connection and disconnection of *devices* from the computer.]
2. You would start the game.
   [You should see one 2D **Game Object** in the foreground, and one 3D **Game Object** in the background. The 2D **Game Object** should appear within a rectangle with a white outline. But it may not be possible to see this outline when the game starts, unless the **Object** was placed next to one

of the four sides, by default. The 3D **Game Object** should appear within a cube, and on top of a square floor. The floor should be made up of an alternating sequence of black and grey parallel lines. The cube should have white outlines. But again, it may not be possible to see these outlines, unless the **Object** was placed next to the sides of the cube, by default.]

3. You would connect the first *Game Controller*.

   [A second 2D **Game Object** should appear.]

4. You would connect the second *Game Controller*.

   [A second 3D **Game Object** should appear.]

5. You would disconnect the first *Game Controller*.

   [The second 2D **Game Object** should disappear.]

6. You would disconnect the second *Game Controller*.

   [The second 3D **Game Object** should disappear.]

7. You would reconnect both *Game Controllers*.

   [Both the second 2D **Game Object** and the second 3D **Game Object** should reappear.]

8. You would move the *analogue device*, on the first *Game Controller*, between both extremes of its vertical axis.

   (If you were using two *analogue devices* instead of one, you would move the first one between both extremes of its axis.)

   [This should move the second 2D **Game Object** up and down, on the screen.]

9. You would move the *analogue device*, on the first *Game Controller*, between both extremes of its horizontal axis.

   (If you were using two *analogue devices* instead of one, you would move the second one between both extremes of its axis.)

   [This should move the second 2D **Game Object** left and right, on the screen.]

10. You would press the first button, on the first *Game Controller*.

    [This should switch the effect of the movement, of the *analogue devices*, to the speed of the 2D **Game Object**.]

11. You would move the *analogue device*, on the first *Game Controller*, up towards the top of its vertical axis. Then you would quickly release the *device* back to its default position.

    [The 2D **Game Object** should start and continue moving up the screen.]

12. You would let the 2D **Game Object** continue moving up the screen.

    [The **Game Object** should stop when it hits the top edge of the hollow rectangle, which forms a boundary around it. You should hear a unique sound when the **Game Object** collides with this boundary.]

13. You would move the *analogue device*, on the first *Game Controller*, down towards the bottom of its vertical axis. Then you would quickly release the *device* back to its default position.

    [The 2D **Game Object** should start and continue moving down the screen.]

14. You would let the 2D **Game Object** continue moving down the screen.

    [The **Game Object** should stop when it hits the bottom edge of the hollow rectangle, which forms a boundary around it. You should hear the same sound, when the **Game Object** collides with the boundary, that you heard when it hit the top edge.]

15. You would tap the *analogue device* upwards, so that the 2D **Game Object** began to move back up the screen. When the **Game Object** had reached the top-half of the boundary, you would tap the *device* downwards so that it stopped.

    [The controlled **Game Object** should be visible halfway up the boundary.]

16. By gently tapping the *analogue device*, on the first *Game Controller*, up, down, left or right, you would force the controlled 2D **Game Object** towards the stationary 2D **Game Object**, on the screen.

    [When the controlled **Game Object** came within close proximity, of the stationary **Object**, a unique sound should be heard.]

17. By gently tapping the *analogue device*, on the first *Game Controller*, up, down, left or right, you would force the controlled 2D **Game Object** into the stationary **Object**, on the screen.

    [When the two **Objects** collided, you should hear a unique sound caused by the collision. When this sound had ended, you should hear a different, second sound immediately following it. The momentum should be conserved during these collisions. So, if the two **Objects** were to collide head on, the stationary **Object** should move off at half the speed of the controlled **Game Object**, since it would have twice the mass. And the controlled **Game Object** should stop. Both **Objects** should eventually come to rest because of friction.]

18. You would repeat step (17) several times, watching the colour of the controlled 2D **Game Object**.

    [The colour of the controlled 2D **Game Object** should have changed from green to either red or blue after step (17). It should then alternate randomly between red and blue after each subsequent collision.]

19. By gently tapping the *analogue device*, on the first *Game Controller*, you would slowly force the uncontrolled 2D **Game Object** towards one of the sides of its enclosure. You would propel it with the controlled **Game Object**.

    [The sides of the surrounding hollow rectangle should be clearly marked by white lines. The uncontrolled **Game Object** should be stationary, just touching one of the sides.]

20. By gently tapping the *analogue device*, on the first *Game Controller*, you would force the controlled **Game Object** into the stationary **Object**, causing it in turn to collide into the boundary.

    [The collision between the **Game Objects** should be briefly audible. But then it should be replaced by the sound of the stationary **Object** colliding with the boundary.]

21. By gently tapping the *analogue device*, on the first *Game Controller*, you would force the controlled **Game Object** into the four sides of the uncontrolled one. The force of the first two impacts would be applied vertically: directly down onto the top, and up onto the bottom of the uncontrolled **Object**. The force of the second two impacts would be

applied horizontally: onto the right and the left of the uncontrolled **Object**.

[When it had collided with the top and the bottom of the uncontrolled **Object**, the **Game Object** should lose all of its momentum to the uncontrolled one. It should come to a complete halt. However, when it had collided with the other two sides of the uncontrolled **Object**, it should only lose part of its momentum. The controlled **Game Object** should continue moving after each collision, having been partially deflected on each side.]

22. By gently tapping the *analogue device*, on the first *Game Controller*, you would move the controlled **Game Object** back to the mid-level of the boundary.

   [The controlled **Game Object** should be visible halfway up the boundary, horizontally adjacent to the uncontrolled and stationary **Game Object**.]

23. You would press the second button, on the first *Game Controller*.

   [This should switch back the effect of the movement, of the *analogue device*, to the position of the 2D **Game Object**.]

24. You would move the *analogue device*, on the first *Game Controller*, between both extremes of its vertical axis.

   [This should move the controlled 2D **Game Object** up and down the boundary.]

25. You would press the third button, on the first *Game Controller*.

   [This should increase the effect of the movement, of the *analogue device*, on the position of the 2D **Game Object**.]

26. You would move the *analogue device*, on the first *Game Controller*, between both extremes of its vertical axis.

   [This should move the controlled 2D **Game Object** up and down the boundary, but over greater distances than before.]

27. You would press the fourth button, on the first *Game Controller*.

   [This should decrease the effect of the movement, of the *analogue device*, on the position of the 2D **Game Object**.]

28. You would move the *analogue device*, on the first *Game Controller*, between both extremes of its vertical axis.

   [This should move the controlled 2D **Game Object** up and down the boundary, but over lesser distances than before.]

29. You would move the *analogue device*, on the second *Game Controller*, between both extremes of its vertical axis.

   (If you were using two *analogue devices* instead of one, you would move the first one between both extremes of its axis.)

   [This should move the 3D **Game Object**, which appeared when the *Game Controller* was connected, backwards and forwards. The movement should be parallel to the black and grey lines, along the floor underneath the **Object**.]

30. You would move the *analogue device*, on the second *Game Controller*, between both extremes of its horizontal axes.

   (If you were using two *analogue devices* instead of one, you would move the second one between both extremes of its axis.)

[This should move the 3D **Game Object** left and right. The movement should be across the black and grey lines underneath the **Game Object**.]

31. You would press the first button, on the second *Game Controller*.

[This should switch the effect of the movement, of the *analogue devices*, to the speed of the 3D **Game Object**.]

32. You would move the *analogue device*, on the second *Game Controller*, up towards the top of its vertical axis. Then you would quickly release the *device* back to its default position.

[The 3D **Game Object** should start and continue moving backwards, along the floor.]

33. You would let the 3D **Game Object** continue moving backwards, along the floor.

[The **Game Object** should stop when it hits the back of the hollow cube, which forms a boundary around it. You should hear the same sound you heard when one of the 2D **Game Objects** hit the hollow rectangle around it.]

34. You would move the *analogue device*, on the second *Game Controller*, down towards the bottom of its vertical axis. Then you would quickly release the *device* back to its default position.

[The 3D **Game Object** should start and continue moving forwards, along the floor.]

35. You would let the 3D **Game Object** continue moving forwards.

[The **Game Object** should stop when it hits the front of the hollow cube, which forms a boundary around it. You should hear the same sound, when the **Object** had collided with the boundary, that you heard when it hit the back of the cube.]

36. You would tap the *analogue device* upwards, so that the 3D **Game Object** began to move backwards. When the **Game Object** had reached the back half of the floor, you would tap the *device* downwards so that it stopped.

[The **Game Object** should be visible halfway down the floor.]

37. By gently tapping the *analogue device*, on the second *Game Controller*, up, down, left or right, you would force the controlled 3D **Game Object** towards the stationary 3D **Game Object** on the screen.

[When the controlled **Game Object** had moved within close proximity, of the stationary **Object**, the same sound should be heard as when the 2D **Game Objects** came within close proximity.]

38. By gently tapping the *analogue device*, on the second *Game Controller*, up, down, left or right, you would force the controlled 3D **Game Object** into the stationary one, on the screen.

[When the two **Game Objects** collided, you should hear the same sound heard when the 2D **Game Objects** collided together. When this sound ends, you should hear the same second sound immediately following it, which was heard when the 2D **Game Objects** collided. The momentum should be conserved during these collisions. So, as with the 2D **Game Objects**, if the two 3D **Game Objects** were to collide head on, the stationary **Object** should move off at half the speed, of the controlled **Object**, since it has twice the mass. And the controlled **Game**

Object should stop. Both **Objects** should eventually come to rest because of friction.]

39. By gently tapping the *analogue device*, on the second *Game Controller*, you would slowly force the uncontrolled 3D **Game Object** towards one of the sides of its enclosure. You would propel it with the controlled **Game Object**.

    [The sides of the surrounding hollow cube should be clearly marked by white lines. And the uncontrolled **Game Object** should be stationary, just touching one side of the underlying floor.]

40. By gently tapping the *analogue device*, on the second *Game Controller*, you would force the controlled 3D **Game Object** into the stationary one, causing it in turn to collide into the sides of the cube.

    [Just as with the 2D **Game Objects**, the collision between the **Game Objects** should be briefly audible. But it should be quickly replaced by the sound of the stationary **Object** colliding into the boundary.]

41. By gently tapping the *analogue device*, on the second *Game Controller*, you would force the controlled **Game Object** into the four vertical sides of the uncontrolled one. The force of the first two impacts would be applied directly forwards into the back, and backwards into the front, of the uncontrolled **Object**. The force of the second two impacts would be applied directly into the left, and into the right, of the uncontrolled **Object**.

    [When it collided with the back and front of the uncontrolled **Object**, the **Game Object** should lose all of its momentum to the uncontrolled one. It should come to a complete halt. However, when it collided with the other two sides of the uncontrolled **Object**, it should produce one of three possible results. The controlled **Game Object** should either jump up in the air or fall back onto the floor. Or it should slowly climb up the invisible ramps, which would be inclined against these two sides, and roll back onto the floor. Or, if it had enough momentum, it should climb up and over the uncontrolled **Object**.]

42. By gently tapping the *analogue device*, on the second *Game Controller*, you would move the controlled 3D **Game Object** to the midway point, between the front and back of the floor.

    [The **Game Object** should be visible halfway down the floor, adjacent to the uncontrolled and stationary **Game Object**.]

43. You would press the second button, on the second *Game Controller*.

    [This should switch back the effect of the movement, of the *analogue device*, to the position of the 3D **Game Object**.]

44. You would move the *analogue device*, on the second *Game Controller*, between both extremes of its vertical axis.

    [This should move the controlled 3D **Game Object** backwards and forwards, along the floor.]

45. You would press the third button, on the second *Game Controller*.

    [This should increase the effect of the movement, of the *analogue device*, on the position of the 3D **Game Object**.]

46. You would move the *analogue device*, on the second *Game Controller*, between both extremes of its vertical axis.

[This should move the controlled 3D **Game Object** backwards and forwards, along the floor, but over greater distances than before.]

47. You would press the fourth button, on the second *Game Controller*.

[This should decrease the effect of the movement, of the *analogue device*, on the position of the 3D **Game Object**.]

48. You would move the *analogue device*, on the second *Game Controller*, between both extremes of its vertical axis.

[This should move the controlled 3D **Game Object** backwards and forwards, along the floor, but over lesser distances than before.]

49. You would press the fifth button, on either the first or the second *Game Controller*.

[This should display a message briefly, stating that the game was about to shut down, before shutting it down. This should also produce a new computer file, next to the **Game Database** used for the test. The new file should be another *Database*, containing one *Record*. This *Record* should be a copy of the **Log Record** produced, in the computer, during the test. This log should contain all the changes that affected the uncontrolled 3D **Game Object**.]

4.1.2 DESIGNING THE SOFTWARE

The task of each **Host Module** has already been outlined in the descriptions of *The Software Architecture*.[3] From these descriptions, you would be able to break down each **Host Module** into a set of *software procedures*.

Once you had decided what *procedures* you would use, each of these would be described in a document. The document would give a name to each one and describe it. The document would also include the description of the **Host Module** which these belonged to. You can copy or refer to the descriptions of the Host Modules in chapter 3 of this book.

Whatever set of *software procedures* you chose, each **Host Module** would need three standard *procedures*. The first *procedure* would be needed to set up the *module*, when the game had begun. A second *procedure* would be needed to periodically update the task of the *module* during each *Unit of game time*. And a third *procedure* would be needed to shut down the **Host Module**, when the game had ended.

As has already been mentioned, when it was set up, the **Central Host** would set up all the other **Host Modules**, beginning with the **Database Host**. This should read, into the computer memory, all the *Records* required by the **Central Host**. So that it could verify each of these *Records* contained sensible values in its *Fields*. If any of these were not sensible, then the **Central Host** would have failed to set up. So it would shut down the game in accordance with its description. Otherwise, it would continue setting up the other **Host Modules**, checking for any errors it detected after each setup.

When it was set up, the **Database Host** would allocate all the space, in the computer memory it would need, from the hardware. Remember that the **Game**

Database would hold all the shared *Game data*. It would also hold all *Abstract data* the **Host Modules** and **Game Objects** would use. The **Database Host** would then load the *Database* into the computer memory. Every *Record* in the *Database* would have sensible values when the *Database* was created. So there would be no need to set up either the *Abstract data* or the *Game data*, after these had been loaded.

When the **Objects Host** was set up, it would find out how many **Game Objects** were going to be used in the initial set, and the size of each one. This information would be found in the **Objects List Record** in the **Game Database**. It would then allocate all the space in the computer memory it would need to store these **Objects**. And each **Object** would, in turn, be created and set up in that space, in accordance with the description of that **Host Module**.

All the other **Host Modules** would simply verify that all the *Records*, to be used by each one had sensible values, when these were set up. Otherwise, it would fail in accordance with the description of the **Central Host**.

In keeping with this description, the **Central Host** would try to update all other **Host Modules**, in turn, when it updated its task. Each of the other **Host Modules** would, likewise, perform whatever task had been outlined in its description, when it was updated by the **Central Host**. The exceptions would be the **Database Host** and the **Objects Host**. These would not require a *procedure* to update the two tasks these would carry out. Since each would perform its task when *Records* were accessed in the **Game Database**, or **Secondary Events** were received from the **Events Host**, respectively.

While performing its task, each **Host Module** would check for any errors, just as when it was set up. And it would report back any errors to the **Central Host**, in accordance with its description. This would include when the **Events Host** received a **Shutdown Event**. It would stop and report this as an error. And this would in turn cause the whole game to be shut down.

Most **Host Modules** would do nothing when each was shut down, apart from the **Database Host**, the **Objects Host** and the **Central Host**. The **Database Host** would release all the space in the computer memory it was using. The **Objects Host** would do likewise. The **Central Host** would shut down all the **Host Modules**, *software libraries*, the computer hardware and close the game.

The 11 **Game Objects** (three 2D **Game Objects**, three 3D **Game Objects**, 2D **Camera Object**, 3D **Camera Object**, **Master Sound Speaker Object**, **Master Player Object**, **Master Physics Object**) which would be used to test the **Event-Database Architecture** have already been described in the previous chapter. All of these **Objects** would use two standard *software procedures*.

The first *procedure* would be used to set up the **Object** with its properties, taken from a *Record* in the **Game Database**. For this test, however, none of the **Objects** need to do anything when set up. Therefore, the *procedure* would do nothing. The second standard *procedure* would be used to respond to the **Events** which would test the **Architecture**.

For example, the controlled 2D **Game Object** would use this second *procedure* to respond to **Secondary Events** from the **Game Controllers Host**. When it received a

Connect Event, the **Object** would add itself onto the list of 2D **Objects** that would be displayed. And when it received a **Disconnect Event**, it would remove itself from the list.

The response of a **Game Object** to **Secondary Events** would depend on its role in the *game design*. It may have only one response. In which case it needs one secondary *procedure* to perform its **Action** in addition to the standard *procedure*. It may have two responses, in which case it needs two secondary *procedures* to perform its two **Actions** in addition to the standard *procedure*. And so on and so on. It may have no response at all. Nevertheless, just like the **Host Modules**, you would include a description of each secondary *procedure* in a *software design* or *technical design* document. The document would give a name to each one and describe it. The document would also include a description of the role of the **Game Object** in the *game design*.

4.1.3 Designing the Database

The design of the test *Database* would have two components: the *Game data* and the *Abstract data*. The shared *Game data* would come from the *data* described in the overall design of the test. These include the artwork and the sounds that would be used. The *Abstract data* would come from the *software design* of the **Host Modules** and the **Game Objects**. These designs, as mentioned in the previous chapter, would include what *data* each **Host Module**, or **Game Object**, would receive, modify or send to other **Host Modules**. So, from these descriptions, you would be able to decide what *Records* would be required in the **Game Database**.

Once you had made your decision, you would need to choose what set of *Database Tables* were going to be in the **Game Database**, *Database Records* in each *Table* and *Database Fields* in each *Record*. Each *Field* could either be a number (floating point number or whole number), a group of words or a *Primary Key* of another *Record*. A *Field* could also be a list of either numbers, groups of words or *Primary Keys*. You would need to select the size of each *Field*. This means the minimum and maximum value of each number, the maximum length of each group of words, the maximum length of each list.

After you have designed the **Game Database**, you should draw an *Entity-Relationship diagram*[4] from it. So that you can visualise it and see the relationship of all the entities or items in the **Database**. This includes the relationship between all the *Database Tables*, *Database Records* or *Database Fields*. You should be able to see where you have any redundant data which you can eliminate and make the **Database** more simple.

Now in order to create an *Entity-Relationship diagram*, it is important to avoid creating a hierarchy of *Database Tables*, *Records* or *Fields*. And avoid creating circular references or relationships between the entities. In the description of some of the **Host Modules**, there were relationships between entities in the **Game Database** that looked like hierarchical relationships.

For example, the **Sounds Host** has a *Database Record*, a **Sound Microphone Object Record**, which refers to another *Record*, a **Game Object Record**, that belongs to the **Objects Host**. And the latter defines a location of a **Game Object** where a microphone in the former has been placed in the *Game World*. To listen to sounds around that locality. This looks like a hierarchical relationship between the **Sound Microphone Object Record** and a **Game Object Record**. With the former one level above the latter in the hierarchy.

But this apparent hierarchical relationship is incidental. The paradigm in creating a *Relational Database* is *Basic Set Theory*[5]: not a hierarchy. In this paradigm, everything is a set which has some relationship with another set. A *Database Table* is a set of *Database Records*. A *Database Record* is a set of *Database Fields*. A *Database Field* is a set of words or numbers. To establish a relationship between one set and another all you need to do is to give them a common member.

In the case of the **Sound Microphone Object Record** and the **Game Object Record**, this common member is the **Object ID Field** of a **Game Object Record**. The **Object ID Field** is a Primary Key of the **Game Object Record**. But instead of being a Primary Key, it could just as well be a *Field* containing a number like

'9999'

or a word like

'MicrophoneObject'.

So long as this common member is unique and can be used to identify a relationship between a **Sound Microphone Object Record** and a **Game Object Record**, then it fits the paradigm. So long as you can use it to find the intersection of one set, a **Sound Microphone Object Record**, and another set, a **Game Object Record**, then it fulfils its purpose.

From this you can see that the **Object ID Field** of a **Game Object Record** is not the ideal common member. It is a Primary Key for a *Record* and therefore unique. But you could have several *Database Records* which have that in a secondary *Field*, not just a **Sound Microphone Object Record**. And if you were to search for all *Records* with the **Object ID Field** with a particular Primary Key, you could get a **Sound Microphone Object Record** or a **Game Controllers Host Record** which also has an **Object ID Field**.

A better common member to have between a **Sound Microphone Object Record** and a **Game Object Record** would be a **SOUND OBJECT FIELD**. This would be a combination of the Primary Key of a **Sound Microphone Object Record** and the Primary Key of a **Game Object Record** e.g.

SoundMic1GameObj1.

And a better common member to have between a **Game Controllers Host Record** and a **Game Object Record** would be a **GAME CONTROLLER OBJECT**

FIELD. This would be a compound of the Primary Keys of a **Game Controllers Host Record** and a **Game Object Record** e.g.

GameCtrl1GameObj1.

By doing this you avoid the pitfalls of creating a hierarchical relationship between the entities i.e. **Sound Microphone Object Record**, **Game Controllers Host Record** and **Game Object Record**. These hierarchical relationships naturally arise in the hierarchical databases used in the Computer Games industry and cause several problems.

Normally in the Computer Games industry, in the *Software Evolution Process* that is used to make a game or *game-editor*, the data for the game ends up in the form of a hierarchical database. This is a natural consequence of the programming languages used to make the game or the *game-editor*, or the commercial *game-engine* that the game or *game-editor* is made from. The paradigm of these programming languages is to treat the Data Structures like branches growing upwards from the trunk of a tree, or branches of a root extending downwards into the ground. In this paradigm, Data Structures are contained within **Game Objects**. Each **Game Object** can either refer to other **Game Objects** in a hierarchy or inherit properties from other **Game Objects** in a hierarchy. And that in turn produces a hierarchy of Data Structures. And that in turn produces a hierarchical database.

But due to the way that the game or *game-editor* evolves through the *Software Evolution Process*, from the beginning to the end of production, without any plan, the hierarchy that results is very deep, very complex and unscrutable. You can get **Game Objects** which inherit from multiple parent **Game Objects**. You can get **Game Objects** which refer to or are inherited by multiple child **Objects** lower down the hierarchy. You can get **Game Objects** lower down the hierarchy referring to or inheriting from **Objects** higher up the hierarchy. That results in a circular dependency which makes it impossible to unravel or visualise. And no one understands the hierarchical databases of these commercial games or *game-engines* as a whole. Here is an example of these hierarchical databases in Figure 4.1.

The hierarchical Data Structure of the commercial *game-engines* does not only affect the ability to visualise the data in single player games. It also affects the ability to visualise the data being transmitted across a computer network in multiplayer games. The data transmitted across the network is used to synchronise the *Game World* on the computers on the network taking part in the game. And this is done by replicating the properties of the **Game Objects** in the *Game Worlds* on these computers. And by replicating the execution of *software procedures* of the **Game Objects**.

Replicating the properties of the **Game Objects** across the network involves replicating the Data Structures of the **Game Objects**. And that Data Structure is part of the hierarchical Data Structure or hierarchical database of the game. So the data transmitted to synchronise these properties is also hierarchical. There is an example

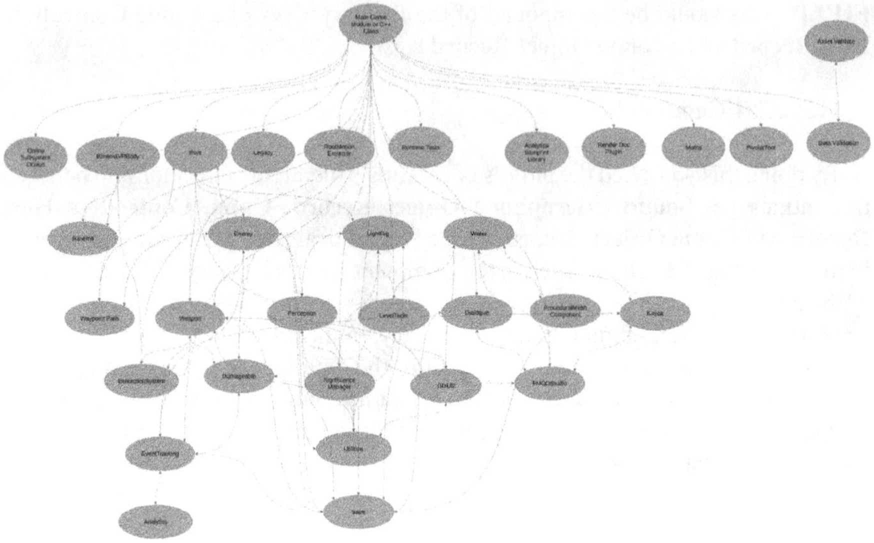

FIGURE 4.1 Visualisation of the hierarchy of Objects or C++ Classes or Data Structures based on inheritance, from a game built with the Unreal Engine by Slippery Games Inc. 2019. Redacted. Anonymous.

of the hierarchical database used to replicate properties on the computer network in Figures 4.2–4.4.

Replicating the execution of the *software procedures* involves replicating the execution of *procedures* which are part of the Data Structures of the **Game Objects**. And the parameters that these *software procedures* take come from the hierarchical

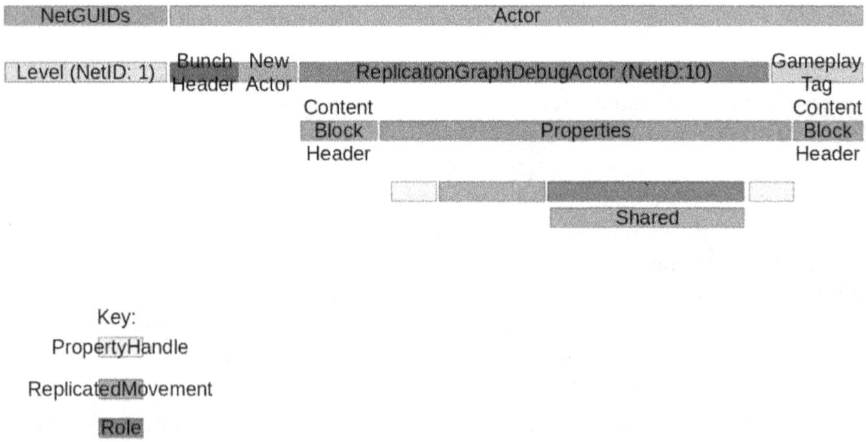

FIGURE 4.2 Visualisation by Unreal Networking Insights Tool of Levels 1, 2, 3, 4 and 5 in the hierarchical Data Structure of a message or packet being transmitted across a computer network to replicate properties of a Game Object.

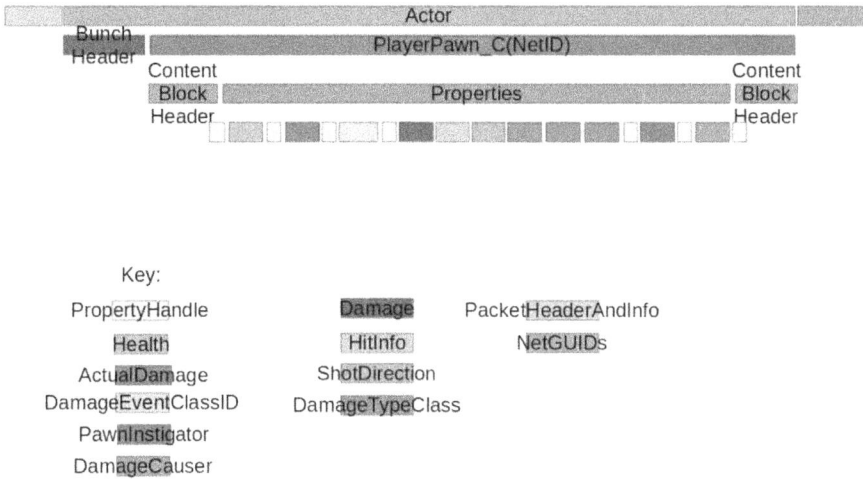

FIGURE 4.3   First half of a visualisation by Unreal Networking Insights Tool of Levels 1, 2, 3 and 4 in the hierarchical Data Structure of a message or packet being transmitted across a computer network to replicate properties of a Game Object (or 'Actor' in the language of the Unreal Engine) called 'PlayerPawn_C'.

Data Structure of the game. So the data transmitted to replicate the execution of these *software procedures* is also hierarchical and very difficult to visualise. These is an example of the hierarchical database used to replicate the execution of *software procedures* in Figures 4.5–4.7.

Note the complexity of the hierarchical Data Structures of the messages in the preceding diagrams. This complexity is not only reflected in the number of levels in

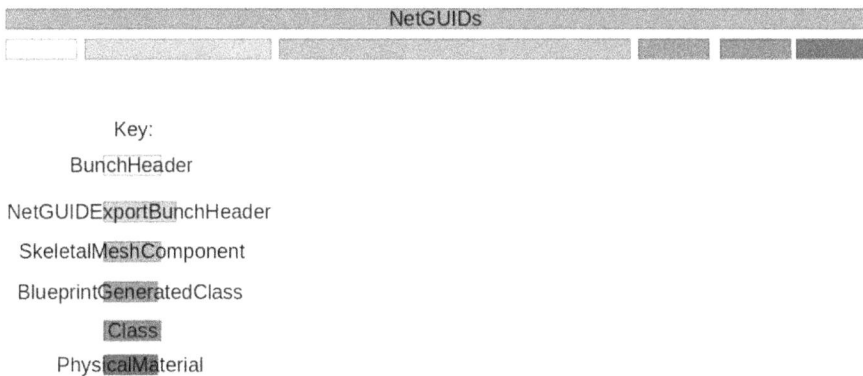

FIGURE 4.4   Second half of the visualisation by Unreal Networking Insights Tool of Level 1 and 2 in the hierarchical Data Structure of a message or packet being transmitted across a computer network to replicate properties of a Game Object (or 'Actor' in the language of the Unreal Engine) called 'PlayerPawn_C'.

PacketHeaderAndInfo	Actor	Actor

FIGURE 4.5 Visualisation of Level 1 in the hierarchical Data Structure of a message or packet being transmitted across a computer network to replicate the execution of a software procedure 'ServerMoveNoBase'.

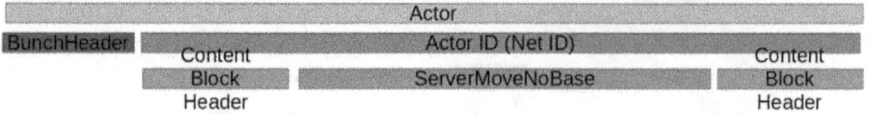

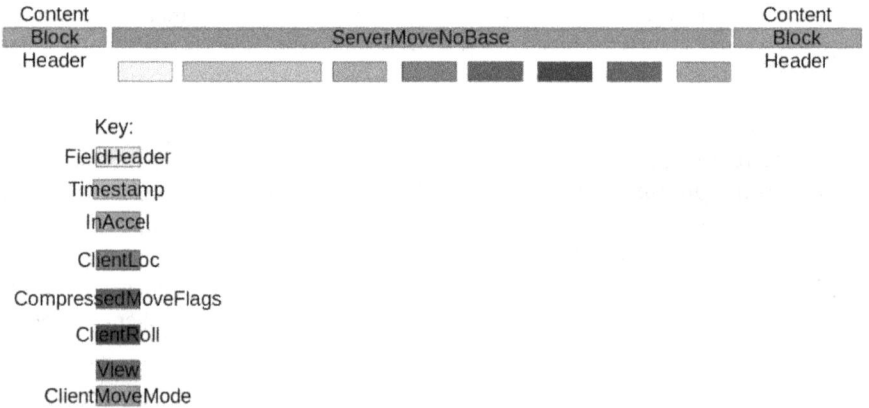

	Actor	
BunchHeader	Actor ID (Net ID)	Content
Content	ServerMoveNoBase	Block
Block		Header
Header		

FIGURE 4.6 Visualisation of Levels 1, 2 and 3 in the hierarchical Data Structure of a message or packet being transmitted across a computer network to replicate the execution of a software procedure 'ServerMoveNoBase'.

Content		Content
Block	ServerMoveNoBase	Block
Header		Header

Key:
FieldHeader
Timestamp
InAccel
ClientLoc
CompressedMoveFlags
ClientRoll
View
ClientMoveMode

FIGURE 4.7 Visualisation of Levels 3 and 4 in the hierarchical Data Structure of a message or packet being transmitted across a computer network to replicate the execution of a software procedure 'ServerMoveNoBase', from a game built with the Unreal Engine.

the hierarchy, and the number of sub-divisions in each level. But it is also reflected in the cryptic names given to each sub-division e.g.

PacketHeaderAndInfo
Actor
NetGUIDs
NetID
GameplayTag
ContentBlockHeader
BunchHeader
NetGUIDExporterBunchHeader

These names come from the degenerative language of the *Software Evolution Process* used to develop the commercial *game-engines* that these Data Structures come from.

Some would say the complexity of the hierarchical Data Structure being used to transmit messages across a computer network, by commercial *game-engines*, simply reflects the complexity of the computer hardware and Network Cards. And it also reflects the low-level techniques used to transmit messages efficiently using this technology (e.g. Binary Bits and Bytes), which only 'experts' can understand. The implication is that this complexity just reflects the state of the art.

But in the end, all these messages do is affect the high-level constructs of the *Game World* i.e. add, move or delete **Game Objects**, display **Objects**, respond to **Events** and perform **Actions**. So why should the names and Data Structures of these messages reflect the low-level technology instead of reflecting the high-level constructs of the *game design*? The technology was created to serve the Users; the Users were not created to serve the technology. The technology does not really care whether the Data Structure being transmitted is hierarchical or not. As has already been explained, the hierarchical Data Structure or hierarchical database for the games built with commercial *game-engines* is incidental. Due to the paradigm of the programming languages used to build *game-engines*, the Data Structures are contained within the **Game Objects**. And the **Game Objects** either refer to each other or inherit from each other in a hierarchy. And that in turn creates a hierarchical Data Structure, based on reference or inheritance. And that in turn produces a hierarchical database. And that in turn produces the hierarchical Data Structure of the messages sent across the network.

If the game were based on a *Relational Database*, then the Data Structure of the messages would reflect that *Relational Database*. Each message would either be referring to an entity in the *Database* (e.g. a **Game Object**, a **Primary Event**, a **Secondary Event**, a *sound stream*, a 2D image, a 3D Model etc.) Or it would be reading the properties of that entity. Or it would be writing to the properties of that entity. So the names of the divisions of the Data Structure would reflect the names of the entities, *Database Tables*, *Records* or *Fields*. These names would not be cryptic and reflect the low-level techniques or tools used to build the game. Since that is precisely the job of the *Database Administrator* who maintains the *Database* in the **Event-Database Architecture**. To keep the language of the *software architecture* reflecting the high, abstract, accessible constructs of the *Game World*, the *game design* and players.

Furthermore, the Data Structure of each message would be a partially ordered set of *Database Fields* which holds the properties of these entities i.e. a pair of values. The first value is the ordinal number of a *Database Field*, and the second value is the value to be read or written to that *Field*. And the first pair of values in each message would be the ordinal number of *Primary Keys*, and the *Primary Key* of a *Database Record*. There is an example of the Data Structure of each message in Figure 4.8.

For example, consider a *Database Table* of **Secondary Events**, which you can see in Table 4.1.

If the **Events Host** wanted to send a message across the computer network to the **Objects Host** to respond to one of the **Secondary Events** in the *Table*, the Data Structure of the message would just be the ordinal number for *Primary Keys* followed

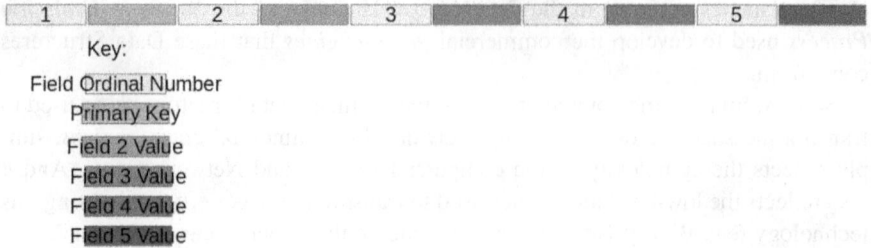

Key:
Field Ordinal Number
    Primary Key
        Field 2 Value
        Field 3 Value
        Field 4 Value
        Field 5 Value

FIGURE 4.8 Visualisation of the pairs of values, the ordinal number of a *Database Field* and the value of that *Field*, that makes up the relational Data Structure of a message or packet being transmitted across a computer network. By a game built with the Event-Database Architecture.

by *Primary Key* for the *Database Record* for that **Event**. There is an example of this in Figure 4.9.

Another example, consider a *Database Table* of **Game Objects** in Table 3.4 described earlier. Now in the **Event-Database Architecture**, the paradigm for making multiplayer games is not to constantly replicate the properties of **Game Objects** or the execution of *software procedures* or functions on computers or **Game Clients**, across a computer network. Unless you are using the **Peer-To-Peer Network**

TABLE 4.1
Example of a *Database Table* of Secondary Events.

Secondary Event ID	Delay (Sec.)	Game Time (Sec.)	Game Object	Causing Objects
Master Object Initial Reset Event	0	1	Master Object	None
Master Object Periodic Reset Event	0	200	Master Object	None
Master Object Heartbeat Event	0	258	Master Object	None
Thief Dead Event	0	260	Thief Object	Warrior 2D Player Object
Thief Resurrect Event	5	265	Thief Object	Thief 2D Player Object

Secondary Event ID	Priority Events ID	Hex. Code
Master Object Initial Reset Event	None	0011
Master Object Periodic Reset Event	None	0018
Master Object Heartbeat Event	None	0021
Thief Dead Event	Thief's Death Priority Events	0022
Thief Resurrect Event	None.	0023

The most important part is the first *Field* the **Primary Keys**. For a full explanation of all the other *Fields*, see the subchapters entitled **3.1 Events Host** and the subchapter entitled **A4.2: Secondary Events Table** in the subchapter **A4 Step 4: LPmud Data Design**.

Key:

Field Ordinal Number

Thief Dead Event

Thief Resurrect EVent

FIGURE 4.9   Visualisation of the relational Data Structure of two messages or packets being transmitted across a computer network, from the Events Host to the Objects Host, for the two Primary Events 'Thief Dead Event' and 'Thief Resurrect Event'.

Architecture. In that case, when a new **Peer** joins a game, the **Central Host** on the special **Peer** hosting the game sends messages to the new **Peer** that connects to the network. To replicate selected *Database Records* or *Database Fields*, once, in its local **Game Database** in order to synchronise the *Game World* on the host and the new **Peer**. And it then relies on the new **Peer** replicating the **Events** and **Actions** that occur on the host, in order to maintain synchronisation.

In that case, to replicate the properties of a **Game Object**, for example, used to move it across the *Game World*, then the Data Structure of the message from the **Central Host**, to the new **Peer**, would just be the ordinal number for Primary Keys. Followed by the Primary Key of the **Game Object**. Followed by the ordinal number for the position of the **Game Object**. Followed by the values for its position. Followed by the ordinal number for the *Database Fields* of its speed. Followed by the values of its speed. Followed by the ordinal number for the *Database Fields* of its acceleration. Followed by the values of its acceleration. There is an example of this Data Structure in Figure 4.10.

When a multiplayer game was started, each **Host Module** would connect to the **Database Host**, and the **Database Host** would send back the ordinal number, size and full name of each *Field* in the **Game Database**. This stream of information would begin and end with the ordinal number of the Primary Keys which is always 1.

Key:

Primary Key Ordinal Number   Angular Position Field Ordinal Number

Thief 2D Player Object   Angular Position Value

X Field Ordinal Number   Angular Speed Field Ordinal Number

X Value   Angular Speed Value

Y Field Ordinal Number

Y Value

X Speed Field Ordinal Number

X Speed Value

Y Speed Field Ordinal Number

Y Speed Value

FIGURE 4.10   Visualisation of the relational Data Structure of a message or packet being transmitted across a computer network, from the **Central Host** to the **Database Host**.

And therefore, each **Host Module** would know the ordinal number and size of any *Field* in any messages it read from other **Host Modules**. And it would know the ordinal number and size of any *Field* in any messages it wrote out to other **Host Modules**. And each **Host Module** could visualise the messages it was receiving or sending, in its logs or on the screen, by displaying the full name of the *Fields* next to the values of each *Field* it was reading or writing across the computer network. Unlike the complex and cryptic hierarchical Data Structures of the messages sent across a computer network, for a game built with a commercial *game-engine*, which is based on a hierarchical database.

In a *Relational Database*, like the **Game Database** of the **Event-Database Architecture**, the **Database** can also be visualised in an *Entity-Relationship diagram*. And this makes it clear for everyone to understand and use. These leads to seven advantages that the **Event-Database Architecture** and its *Relational Database* have over commercial *game-editors* or *game-engines* and their hierarchical databases:

1. The hierarchies in the database of commercial *game-engines* can become so complex that these cannot be visualised in a diagram like the **Game Database** of the **Event-Database Architecture** can in an *Entity-Relationship diagram*.
2. The hierarchies in the commercial *game-engines* can become circular, where a Data Structure lower down the hierarchy refers to one further up the hierarchy, causing circular references in the hierarchy, whereas the **Game Database** of the **Event-Database Architecture** does not have circular references if developed within the paradigm of *Set Theory*.
3. Modifying one Data Structure in the hierarchy of the commercial *game-engines* may require modifying some other Data Structure further down the hierarchy related to it, but the hierarchies are so complex no one can predict this, and so you suddenly get corruption in the hierarchical database when it is rebuilt, because two interrelated Data Structures were not rebuilt, whereas the **Game Database** of the **Event-Database Architecture** is a *Relational Database* where the Data Structures are connected by members they share in common, not by a hierarchy.
4. The sudden corruption of the hierarchical database of commercial *game-engines* makes the process of building the game non-deterministic; therefore, you often have to rebuild the entire database from scratch to stop this and make the process deterministic, which waste a lot of time and resources.
5. The hierarchical Data Structures of commercial *game-engines* lead to complex and cryptic hierarchical Data Structures of the messages sent across a computer network when playing multiplayer games.
6. The hierarchical databases of commercial *game-engines* cannot be queried and edited using a standard tool, so there are no books available which describe how to query and edit the entire database.
7. A *Relational Database* can be queried and edited using standard programming language called Structured Query Language or *SQL*, and there are millions of books available which describe how to query and edit the entire database.

An example of point (2) can be seen in some commercial *game-engines* that store the properties of **Game Objects** in files. The properties of each **Game Object** are stored in one file. So when the game is **SAVED** or written to a file to continue playing later on, all of the properties of the **Game Objects** in the *Game World* are **saved** to files. And when the game is **LOADED** or read back from a file, to continue playing, all of the properties of the **Game Objects** in the *Game World* are **loaded** back from other files.

However, when you have a hierarchical relationship between the **Game Objects**, where one **Game Object** higher up the hierarchy refers to or is inherited by another **Game Object** lower down the hierarchy, that means you have a hierarchical relationship between the files for the **Game Objects**. And this leads to common *Bugs* in these commercial *game-engines* when you have a circular relationship in this hierarchy, which results in the process of **loading** the game never ending.

For example, suppose you try to load a game, which has three **Game Objects** in the *Game World* that have a circular relationship in a hierarchy. You begin by trying to load the properties of the first **Game Object** that includes a reference to the second **Game Object** at the next level down in the hierarchy. So you have to load the properties of that second **Game Object** from a second file. And when you **load** the properties of the second **Game Object** from the second file, you find a reference to the third **Game Object** at the next level down the hierarchy. So you have to **load** the properties of the third **Game Object** from a third file. But when you **load** the properties of the third **Game Object** from the third file, you find a reference to the first **Game Object** from the first file, two levels further up the hierarchy. So you have to load the properties of the first **Game Object** again from the first file, and so on and so on.

Before you know it, the process of loading the **Game Objects** gets stuck in a circular loop that never ends. In this example, it is obvious only because the hierarchy of **Game Objects** is only three levels deep. And it may seem easy to prevent. But imagine where the hierarchy is 20 or 30 levels deep. At that point it is not obvious and it is almost impossible to prevent. You will find yourself working on a game with a commercial *game-engine* for a long time, seemingly without any problems, and then suddenly the game stops working. And it seems to be stuck in the process of **loading** a previously **saved** game. Leaving you completely baffled.

Now some commercial *game-engines* do report warnings or errors when they detect this happening. But these reports are just remedial and flawed.

Firstly, these reports are inconsistent and do not always appear when playing a game. Due to the fact that the symptoms of this circular never-ending process of **loading** sometimes cause larger overriding problems like running out of computer memory.

Secondly, these reports are vague. The reports may tell you that you have a circular dependency in your **Game Objects**. But these will not tell you what all the **Game Objects** are that make up this circular dependency in the hierarchy.

Thirdly, these reports are incidental. These reports will only tell you of the circular dependencies you happen to come across while playing the game. It will not tell you of all the other circular dependencies that may exist in the hierarchical database.

Fourthly, these reports are ad hoc. These reports appear after the game has been built and the hierarchical database is being used to play the game. These reports do appear before the game is built. And as a result these reports waste a lot of time and resources. You get none of these four problems with a Relational Database.

4.1.4 BACKGROUND RESEARCH

Before you implemented the **Event-Database Architecture**, it would be worth-while doing background research on what techniques could be used to build some of the **Host Modules**. These would namely be the **Database Host**, **Physics Host**, **Graphics Host** and **Sounds Host**.

For the **Database Host**, you could consider using *Hash-Tables*.[6] These allow you to quickly look up a table of information, given an *ID* for an entry. Each entry is placed in the table using its *ID*. So you could use the Primary Key of a *Record* to find or set its position (i.e. first row, second row and third row) in a *Database*. Since the *ID* of an entry, in a *Hash-Table*, could be a number or a word, the Primary Key could also be a number or a word. There are various *Professional Database sources*[7] covering how to store and retrieve *Records* and *Fields* in a *Database*. And there are also various *Data structure sources*[8] describing how to create *Hash-Tables*.

For the **Physics Host**, you could consider reading about how the laws of Physics could be modelled in software. You could consider how the position, speed, acceleration, direction and mass of **Game Objects** would be represented in the *Records*, of the **Game Database**. One common method uses mathematical *Vectors*. This may involve either breaking the position, speed and acceleration of each **Object** into two or three geometric components. So each quantity would have an X, Y or Z coordinate. Or it may simply involve breaking each quantity into just two components: one representing the direction of the *Vector*, and the other its size. There are many *Applied mathematics sources*[9] that explain how *Vectors* may be used to model physical quantities.

For detecting the collision between **Game Objects**, you could consider reading about how you would detect the overlap of different volume shapes. These would include the overlap of any two 2D shapes: such as a square, a rectangle, a triangle or a circle. These would also include the overlap of any two 3D shapes: such as a cube, a cylinder, a cone or a sphere. It would be useful to understand how you would detect when a point was contained within a 2D *polygon*, when two lines intersected in 2D or 3D space, and when a line intersected with a 3D *polygon*.

For determining the angle of reflection, after a collision, you could again use mathematical *Vectors*. If the momentum of each **Object** were represented by a *Vector*, then you would be able to handle each collision simply. So when two mobile **Game Objects** collided, you could transfer the *Vector*, from one **Object** to the other, to produce the resultant angle of reflection. Similarly, when a mobile **Game Object** rebounded off the surface of another **Object**, *Vectors* may be used. A common method involves using the impact *Vector* of the **Object**, and the *Normal Vector* of the surface, to calculate the resultant *Vector* after collision. The surface may be one side of a 2D shape, or one face of a 3D *polygon*.

In order to stop static items from moving, after collision, some Computer Games simply do not apply any momentum to the item. Instead, the game preserves the momentum of the moving body, when it rebounds off the surface of a static item. Or the game reduces the momentum by an amount proportional to the speed of the moving body. There are several *Physics sources*[10] that cover how the motion of physical bodies may be simulated in a Computer Game.

For the **Graphics Host**, you could consider studying how the 2D shapes or the 3D models would be projected from 2D or 3D space, onto a computer screen. This usually involves using mathematical Matrices. You could use Matrices to determine the position of each shape from a viewpoint, in 2D or 3D space. You could use Matrices to project the *vertices*, of the 3D models, through the viewpoint, onto the 2D space of the computer screen. You could also use Matrices to move and rotate the shapes or the models. There are many *Mathematics sources*[11] that discuss how Matrices may be used to perform projections, and to move 2D shapes or 3D models.

To draw 2D images, or 3D models, it would be helpful to understand how the *polygons* would be filled in using *Textures* and *Texture coordinates*. You could also read about how the *data* for the *vertices*, *Textures* and *Textures coordinates* could be stored in the *Database*. There are numerous *Computer graphics sources*[12] that describe how 2D images, and 3D models, are displayed on screen, and how to achieve visual effects.

Finally, for the **Sounds Host** you may find it useful reading about Digital Signal Processing. This is a large topic. But only a few methods within it are necessary for creating a reasonably broad range of sound effects that could be used in a Computer Game. These include methods that reverberate a sound, filter out different frequencies of a sound or distort the sound. These also include methods that mix or concatenate two sounds together. It would also be helpful to find out the different ways in which a *sound stream* could be encoded and stored in the *Database*. There are several *Sound engineering sources*[13] that cover Digital Signal Processing.

4.1.5 DOCUMENTATION TOOLS

An important part of the feasibility test would be the tools which can be used to produce documentation for the test. These include the tools which the *Game Programmers* would need to plan the software they would write, tools the *Game Artists* would need to plan their artwork, tools the *Game Designers* would need to plan their changes to the *game design*, tools the *Sound Designers* would need to plan their sounds and music, and tools the *Game Testers* would need to plan their execution of the test and generating a report.

Explicitly documenting the different designs you produce would have three advantages. Firstly, it would make it easy for you to refer to these designs to implement any component of the software. This may be a *software module*, a piece of art, a piece of sound or the **Game Database**. Secondly, the documentation would help other people understand the components, if they were required to maintain it, or improve it. The documentation would help them understand the inter-dependencies between the components. Finally, the documentation could be used to test each component once it had been built.

To produce these documents, you would need tools that enabled you to write simple articles and check the spelling. You would be able to highlight words, with different fonts, so that you could help the reader recognise and differentiate between passages and topics. You may also be able to draw simple annotated diagrams, flow-charts or tables, with these tools. This would provide you with a concise description of any designs you created.

Either many of the documents you produced would be interrelated. Or many parts of each document would be interrelated. This would be because the components described by each document, or each part of a document, were interrelated.

For example, the menus described in a *game design* would be interrelated. Also the *software modules* of a *technical design* would be interrelated. So after changing any one document, or any one part of a document, you may need to change other documents, or other parts, as well.

It would be very useful, but not necessary, to find a tool that could help you do this automatically. The tool would allow you to link documents, or different parts of one document together. So that when you changed a document, or part of one, it would either list all the other documents, or other parts which had to be changed. Or it would automatically perform these changes somehow. Perhaps, the tool would allow you to describe the relationships between the documents, or different parts. And it would use this description to perform each change.

For example, the tool may allow you to link each occurrence of the name of a certain topic, mentioned in a document. So that when you change the name in one part of the document, all other occurrences of that name would automatically be changed as well. Or the tool would keep a record of the position in the document, of all the unmodified names, which you could study at a later date.

There are many commercial software tools that would provide you with some or all of these capabilities. There are also *Free software*[14] ones, such as OpenOffice (https://www.openoffice.org/). There are various *Electronic documentation sources*[15] that provide a useful introduction to this software.

4.1.6 PROGRAMMING TOOLS

You could use whatever programming language you wanted to write each **Host Module** or **Game Object**. Typically, most people would choose one which could be used across different computer hardware. Two common favourites are the programming languages 'C' or 'C++'. You could also use Python Script, Javascript, Java, WebGL, HTML, Perl.

There are many written *Programming sources*[16] on these two languages. There are many commercial tools which could be used to write software in 'C' or 'C++'. There are also *Free software* ones, such as the GNU C Compiler or GCC. There are several *Compiler sources*[17] that explain how to use GCC.

It would be worth investing in *Revision Control Software*[18] too. This could be used to keep an archive of the computer files used to build the game. This would give you the option of going back to previous versions of the software, if the latest version were to become corrupted, discarded or lost. You could also use this tool to compare and merge differences in any two versions of the game, or a given *software module*.

Or any two computers files containing artwork or sound (provided these files were in data format based on text).

There are numerous commercial tools that could keep this archive. There are also *Free software* tools, such as the Concurrent Versions System or CVS. There are a few *Revision Control sources*[19] that describe CVS and its uses.

4.1.7 SOFTWARE LIBRARIES

The **Events Host**, **Objects Host** and **Physics Host** would not require third-party *software libraries*. You could implement these using only a programming language, and the background research carried out before starting to implement the **Event-Database Architecture**.

However, the **Database Host** would require a *File library*. This would enable it to access whatever storage media the computer hardware had. The **Game Database** would be stored on this media and loaded into the computer memory by the **Database Host**, when it was set up. There is a standard *File library* that several computer hardware use. This would be described in the *Programming sources* that you may read, and other *File library sources*.[20] Although this *File library* was meant to provide a standard *Interface* for any device connected to a computer, hardware vendors typically limit its use to some storage media, and not for others. Or there are subtle variations of the *library*, or extensions added to it. So each hardware vendor provides their own version of the *library*, and any other *software libraries* required for their storage media. These come with the proprietary tools they supply with their proprietary hardware, for developing software on it. And these tools are accompanied by the *User Manuals* describing the *libraries*.

Alternatively, instead of attempting to create the **Database Host** from scratch, you could use a *software library,* created by a third party for hosting a professional *Relational Database Management System* or RDBMS. This would allow you to store any *Relational Database*, including the **Game Database** partly in the computer memory and partly on the storage media. This would allow you to quickly read, write and query *Database Tables, Database Records and Database Fields*. Ideally, it should be small and efficient enough to run on computer hardware with limited resources, which is more often than not the case with hardware that runs Computer Games. You can read about this in various *RDBMS resources*.[21]

The **Graphics Host** would need a *Graphics library* to display 2D images and 3D models on the screen. Although each vendor, again, provides their own, different *library* for their hardware, a standard *library* is available. This is called the Open Graphics Library or OpenGL®. It has an open *Interface*, which is available for many computer hardware. It automatically uses many of the techniques you would come across doing background research. So you would find that it may not be necessary to implement much of the techniques, for displaying 2D images or 3D models, yourself. The *software library* already does this for you. There are several *Graphic library sources*[22] that cover OpenGL.

Although the *Graphics library* may well perform much of the mathematical techniques used to display images, it may omit others that you might want to use elsewhere. These would namely be those techniques, involving *Vectors* and Matrices, you would use to build the **Physics Host**. For this reason, you may find it useful to acquire a *Mathematics library* for both the **Physics Host** and the **Graphics Host**. There is a standard *library* available, that you may come across when reading the *Programming sources*. But it does not include the more advanced techniques you may want to use.

The **Sounds Host** would need a *Sound library* to play sound and music, on the computer hardware. There is no standard *library* available. Each hardware vendor provides their own. However, there have been recent attempts to establish a standard. This is called the Open Audio Library or OpenAL®. There are a few *Audio library sources*[23] that discuss OpenAL.

Finally, the **Game Controllers Host** would require a *Game controller library*. Again, there is no standard *library* available. Each hardware vendor provides their own.

Nevertheless, there may be a *software library* available, for some computer hardware, that combine the ability to interact with *Game Controllers* and play games, with other forms of entertainment. These include playing videos or music, on the hardware. There are commercial versions of such *libraries*. And there are also *Free software* ones, such as the Simple DirectMedia Layer or SDL. There are a couple of *Multimedia library sources*[24] that cover SDL and how you could use it.

Whenever a hardware vendor does not provide a *software library*, or there is no standard *library*, at hand to easily build a **Host Module**, it merely shifts the burden. You would then have to write that *software library* yourself, before you built your Computer Game.

4.1.8   ART TOOLS

To create 3D models, you would need a 3D modelling software. This would allow you to create models using a mesh of *polygons*. It would allow you to fill in the *polygons*, using whatever *Textures* you choose. You would be able to view each model before exporting its *vertices*, *Textures* and *Texture coordinates* to a computer file. The *data* from that file would then be added to a *Record*, in the **Game Database**.

The modelling software would also allow you to create 3D animations. Either you could set the pose of the whole 3D model, for each *Frame*. Or you could just use the skeleton of the model. You would be able to specify the different positions of the skeleton, in each *Frame* of the animation. After watching a preview of the result, you would then be able to export the positions of the model, or its bones, in each *Frame*, to a computer file. This *data* could then also be added to a *Record* in the **Game Database**.

There are several commercial software available which give you all these features. There is also a *Free software* package you could use, that is available for many different computer hardware, called Blender®. There are good *Computer Aided Design sources*[25] that describe these tools.

To create 2D images, you would require different software. You would need a Digital imaging software. This would allow you to create any image, or set of images in an animation sequence, using whatever colours you choose. You would either be able to use a colour palette, and a set of colour indices, to draw the pixels of the image. Or you would be able to create the image by specifying the individual red, green and blue colour components of each pixel. The former method saves more space, in the computer memory, than the latter. The choice of method would depend on how much detail you wanted there to be in the image. But the methods available may depend on the computer hardware the game was for. Some hardware only permit one method.

The Digital imaging software would allow you to export the size of each image, its colours and any accompanying colour palette, to a file. You would then be able to add that *data* to a *Record* in the *Database*.

There are several commercial Digital imaging software available. There is also a *Free software* package you could use called the GNU Image Manipulation Program or GIMP (https://www.gimp.org/). There are numerous *Digital imaging sources*[26] that cover these tools.

4.1.9   DATABASE TOOLS

There are five sets of tools which would be required to manage the **Game Database**. The first set would be used to convert the *data*, from the computer files produced by any third-party software. This third-party software may be, for example, used to create 3D models. The *data* would be converted, into the *data format* used by the *Database*, to create new *Records*. The second set of tools would be used to add these new *Records*, from the converted files to the *Database*. The third set would be used to view and edit the *Records* in the *Database*. The fourth set would be used to extract the *Records*, from the *Database*, into computer files. The fifth set would be used to convert the extracted *Records*, from these files, back into the *data formats* supported by the third-party software.

The last set of tools would allow you to reverse the process of the first set. Similarly, the fourth set would allow you to reverse the process of the second set. Being able to reverse the process would give you the same advantages as using *Revision Control Software*. This would namely be the confidence to correct errors.

For example, it would give you the flexibility to work with the *Database Records* in whatever *data format* was best suited to viewing or editing that *data*. Viewing *Records* which held 3D models, or 2D images, would be best done with the software tools that were used to create these models and images.

Another advantage of being able to reverse the process would be that you could keep the *data* in an intermediary format. This would be useful when the *data formats* changed.

For example, suppose the format of some of the *Database Records* was about to be changed. You could use the fourth set of tools to extract these *Records*, into computer files. After the *data formats* were changed, you would modify the second set of tools to add the extracted *Records*, back into the *Database*. The new tools would add any new *Fields*, to the *Records*, with sensible default values. And the tools would omit any *Fields* that were no longer required.

A reversible process would have yet more advantages. You would not need to keep an archive of all the computer files which were used to build a version of the *Database*. You would only need to keep an archive of the *Database* and the tools. If you wanted to go back to work on a past version of the game, you would get its

Database and its tools from the archive. And you could use these old tools to recreate the old computer files.

There are many commercial *Database* tools available, which could be used as the third set of tools. There are also some *Free software* ones available, such as MySQL® (https://www.mysql.com/), which could also be used. There are various *Relational Database management sources*[27] that discuss how to use MySQL®.

There may be some commercial *Database* tools which combine all five sets of tools in one. But typically, you would have to write your own software tools for the first, second, fourth and fifth sets. To do this, either you would need to understand the *data formats* of the files exported by third-party software. Or you would need to find out how you could control the *data formats* of the files exported. These information would normally be explained in the *User Manuals* of these third-party software.

4.1.10   OPEN DATA FORMAT

You would have to choose an *open data format* for describing the *Database Tables, Records and Fields* of the **Game Database**. The advantages of using an *open data format* have already been outlined in Chapter 2.

Another advantage of using an *open data format* for this is that it allows you to merge one branch of changes to **Game Database** with another branch. Often in the Computer Games industry, a Software Repository is used to keep track of changes made to the files used to build the game. A Software Repository is a recorded archive of changes made to the files used to build game. It allows you to monitor the amount of changes going in, recover any lost changes and track who is responsible for changes and when these changes were made. This is very useful since often, as already mentioned, the Software Production Process defaults and degenerates into a *Software Evolution Process* with lots of changes being made rapidly, with little or no records being kept. A Software Repository helps you mitigate this to a limited extent by keeping some records of the changes. When the Users submit changes to the Repository, they normally have to include one line describing those changes. It is not ideal, but it is better than nothing.

However, the limitation of this is when there are so many changes going into the game, which often conflict with each other, the process of building the latest version of the game, from the Software Repository, becomes very unstable.

At this point, you can mitigate the situation by taking a copy of the files in the main branch of the Software Repository, and putting them in another Repository, which is called a Child Branch. You can then assign a subset of the staff to work on a new feature in this Child Branch. This reduces the number of staff working on the main branch and helps stabilise the situation. After that new feature has been completed, you can then merge the changes in the Child Branch back to the parent branch or main branch.

Nevertheless, this is not trivial especially with commercial *game-engines*. The **Game Database** of commercial *game-engines* is primarily made up of files encoded in a closed proprietary format. Especially, the 2D images and 3D models created by the *Game Artists*. That means that when you branch the files used to build the game and make a lot of changes to the branch, you cannot merge the results back to the parent or main branch. Especially when those changes include changes to the artwork.

But when the entire **Game Database** is stored in an *open data format*, you can merge the branches. The **Game Database** is interoperable with other tools. That means you can create tools to merge **Game Database** in one branch with another branch. Or the software typically used to manage the Software Repository can do this for you. So long as the *open data format* is a Text Format based on *ASCII*[28] characters, as supposed to a Binary Format.

One example of a format you can use to store 2D images is *SVG Format*.[29] One example of a format you can use to store 3D models is *FBX Format*.[30] The latter format comes in two forms. One is a Binary Format. And the other is a Text Format. You can use the Text Format. Both *SVG Format* and *FBX Format* are only suitable for 2D images and 3D models. You cannot store an entire *Database* in these two formats. So, for the rest of the **Game Database**, you will have to use another *data format*.

One such *data format* which you can use for the rest of the *Database* is the Extensible Markup Language or *XML Format*.[31] But, unfortunately, there are no agreed standards on how to use these to describe *Databases*. So this means that for creating a large, central *Database*, and being able to edit it using *Database* software, you would not be able to use these *data formats*. For example, *Relational Database Management Systems* that support *XML Format* are IBM DB2, Microsoft SQL Server, Oracle Database and PostgreSQL.

There are also other *open data formats* such as *JSON Format*[32] that could be used to describe a *Database*. But *JSON* describes *Database* or Data Structures in a hierarchical way or hierarchical model. Each *Database Table* can have a variable number of attributes or *Database Fields*. Each *Database Field* in turn can either

- be a value of any size and any type (i.e. a word or a number or collection of words and numbers); or
- be a variable list of values of any size and any type; or
- be a parent of a variable list of other *Database Fields*; or
- be a parent of another *Database Table*.

And these children in turn can be parents of other *Database Fields* or other *Database Tables*. And so on and so on in a hierarchy. There is an example of how information about the quests available in a game would be stored in *JSON Format* in Figure 4.11.

This hierarchical model conflicts with a relational model used by *Relational Databases*, such as the **Game Database** of the **Event-Database Architecture**. In a relational model, there is a fixed number of attributes or *Database Fields* in each *Database Table*. And each *Database Field* in turn can either

- only be a value of fixed size and one type (i.e. only words or only numbers or only list of words and numbers); or
- only be a list of values of fixed size and one type.

```
{
    player: "Bilbo",
    quests:
        [
            {
                name: "Mount Doom",
                reward: "The One Ring",
                experience: 1000000
            },
            {
                name: "Mines of Moria",
                reward: [
                        "golden swords",
                        "golden chalice",
                        "golden sceptre",
                        "golden ring",
                        "golden crown"
                    ]
                experience: 2000000
            },
            {
                name: "Mirkwood",
                reward: "Boots of Radagast",
                experience: 5000000
            }
        ]
}
```

FIGURE 4.11   A table of quests available to a player in hierarchical model of a *Database* in JSON Format.

A *Database Field* cannot be another *Database Table* as shown in Figure 4.12.

One solution to this conflict is to commit to a hierarchical model and use a *Database* that is only based on *JSON Format* such as a NoSQL *Database* like Oracle NoSQL Database, or Azure Cosmos DB. You just have to be careful that you do not create a hierarchy i.e. you never have a *Database Field* that is a parent of other *Database Fields* or other *Database Tables*. To ensure that the *Database* still matches the relational model of the **Game Database** of the **Event-Database Architecture**.

The advantage of a NoSQL Database is that you can create very flexible *Database Tables*. With very flexible attributes or *Database Fields* which can hold values of any type of *data*. This can be useful when you are unsure about the exact Data Structure you need for your game.

The disadvantage of a NoSQL *Database* is, as the name suggests, you cannot use the standard programming language for querying *Databases* which is the Structured Query Language or SQL. Each one has a non-standard programming language for querying it and therefore has fewer books available to explain how to use it.

Another solution to this conflict is to have hybrid model. That is a *Database* that is based both on a hierarchical model and a relational model. The part of the *Database* based on relational model is in *CSV Format*.[33] And part based on the hierarchical model is in *JSON Format*. An example of *Database* which uses a hybrid model is MariaDB.

Quest ID,	Reward,		Experience,	Player
		quests_csv_format		
Mount Doom,	The One Ring,		1000000,	Bilbo
Mines of Moria,	golden sword; golden chalice; golden sceptre; golden ring; golden crown,		2000000,	Bilbo
Mirkwood,	Boots of Radagast,		5000000,	Bilbo

FIGURE 4.12 A table of quests available to a player in a relational model of a Database in a Comma Separated Format or CSV Format.

Another solution is to commit to a relational model based solely on comma-separated format or *CSV Format* and *ASCII* characters. Many *Relational Database Management Systems* support this format, with minor variations. Examples of *Database* which use a relational model and a *CSV Format* are Oracle DB and MySQL DB. But the problem with embedding Data Structures in JSON Format, in other Data Structures in CSV Format, is that there is a conflict between the Formats which can lead to confusion. Both Formats use the comma character, to separate Fields. And both enclose the names and values of Fields in double-quotation marks. This leads to confusion when you come across a comma or double-quotation mark in a middle of a Database Record in CSV Format, which itself contains a Field whose value is a Data Structure in JSON Format. Is the comma part of the JSON Format embedded in the Field? Or is it part of the Database Record that contains that Field, in CSV Format?

In *CSV Format*, the entire *Database* would be described by one table made up of *ASCII* characters. The *data* in each *Field* would be made up of *ASCII* characters. Each *Field*, in a *Record*, would be separated by a comma (,) character. And each *Record* would terminated by a *Newline character*.[34] The number of columns in the table would be determined by the longest *Record* in the *Database*. Each *Record* would use as many *Fields* as it required, from the left to the right column of the table. Any excess *Fields*, which it did not use, would be left empty and would not contain any *ASCII* characters.

When using the *CSV Format* to describe the **Game Database**, the first *Field* in each *Database Table* would be Primary Keys. These should be unique and could be a number or a word. The rest of the *Fields* would hold whatever *data* that *Table* required.

The first *Database Table* in the **Game Database** would give a brief description of all the other *Database Tables*. The first *Field* of this *Table* would of course be the Primary Keys. These would be unique names for each of the other *Tables* in the **Game Database**.

The second *Field* would be used to give a brief description of the *Records* in each *Table*. The description could just be the name of the *Abstract data* contained in the *Table* e.g.

Primary Events Table
Secondary Events Table

Or it could be the name of the encoded *data* contained within the *Table* e.g.

Sound Stream Table
Animated Vertices Table
3D Models Table

Or it could be a description of the *Fields* in the *Record* e.g.

2D Polygon ID;2D Vertices;2D Normals
3D Model ID;3D Vertices;3D Normals

Either way, the description would not be used as a substitute for a full description of *Database Tables* in some kind of *data design*.

In order to create *Fields* which contained a comma (,) or a *Newline character*[35], a special *Escape character*[35] would be used. Whenever this character was encountered, it would mean that the next character was part of the current *Field*. If the next character were the *Escape character*, then that character would also be part of the current *Field*. So, for example, to include the *Newline character* in a *Field*, you would simply precede it with the *Escape character*. Similarly, to include a comma in a *Field*, you would also precede it with another *Escape character*. For the *Escape character*, most *Database* software use the back slash (\) character.

However, some *Database* software do not recognise *Escape characters*. This is the minor variation in how these software support the *CSV Format* or comma-separated *ASCII* format. To accommodate these variations, each comma or *Newline character*, in a *Field*, would have to be represented by a special sequence of characters. You could use a sequence which begins with the *Escape character* followed by the *ASCII* code number for a comma or a *Newline character*. Or you could use something else. Either way, it would not matter if the *Database* software ignored the *Escape character*. So long as the game and all of its tools could identify the *Fields* in each *Record* and knew what the special sequences were, then the integrity of the *Database* would be kept intact. This would be provided, of course, that neither the game nor the tools altered these sequences.

Thus, you could use the *CSV Format* to construct a computer file for your **Game Database**. When the **Database Host** loaded each *Record*, from this file, the **Database Host** would recognise the type of *data* stored in each *Field* by the characters used within it. Remember that each *Field* could either be a number, a group of words or a Primary Key of another *Record*. A *Field* could also be a list of either numbers, groups of words or *Primary Keys*.

If a *Field* were a number, it would be made up of a single word of numerical *ASCII* characters, along with one optional minus (−) or decimal point (.). If such a *Field* contained a decimal point, then it would be a floating point number, as supposed to a whole number.

If a *Field* were a group of words, it would contain any readable, alphanumerical *ASCII* characters. This means any alphabetical, numerical or punctuation characters. Furthermore, it would contain at least two characters, and one of the characters would not be a numerical character.

If a *Field* were a Primary Key, it would begin with a readable, non-alphanumerical *ASCII* character. This means a character like hash (#), ampersand (&), asterisk (*), dollar ($) or pound (£). This character would be followed by either numerical characters or alphanumerical characters. This would depend on whether you were using numbers or words as Primary Keys. The non-alphanumeric character at the start of the *Field* would, however, be ignored when the *data* was read into the computer memory. It would only be used to indicate that the *Field* was a Primary Key.

If a *Field* were a list of numbers, groups of words or Primary Keys, it would contain two or more entries. For the sake of consistency, each entry in the list would be separated using the same principle used to separate *Fields*, in *Records*. This means each entry would be separated by a comma (,) character, after the *Field* had been extracted from the *Record*. Therefore, before the *Field* was extracted from the *Record*, each entry should be separated by an *Escape character* followed by a comma (,) character.

Similarly, if a comma (,) were meant to be part of an entry in a list, then it would be preceded by an *Escape character*, after the *Field* had been extracted from a *Record*. This means, before the *Field* was extracted, the comma (,) should be preceded by three *Escape characters*.

If a *Field* were meant to be a list of numbers, then all the entries in the list would be like a single numeric *Field*. That is to say, each entry would be made up of only numerical characters, with one optional minus (−) or decimal point (.) character.

Similarly, if the *Field* were meant to be a list of groups of words, then all the entries in the list would be like a single alphanumeric *Field*. That is to say, each entry would be made of at least two alphanumerical characters, one of which was not a numerical character.

Likewise, if the *Field* were meant to be a list of *Primary Keys*, then all the entries would be like a *Field* that contained a single Primary Key. That is to say, either each entry would begin with one non-alphanumerical character, followed by numerical characters. Or each entry would begin with one non-alphanumerical character, followed by alphabetical characters. Again, this would depend on whether numbers or words were being used as *Primary Keys*. Furthermore, the non-alphanumerical characters would be ignored when determining each number or word in the list.

Any list that was meant to be empty, when the game began, and filled in as it progressed, would not be left empty. Instead, space would be reserved in the *Field*, for this list. This would enable the **Database Host** to know how much space, of the computer memory, to reserve for the entire *Database*. So the *Field* would either be filled with a list of zeros (0), if it were going to hold a list of numbers. Or it would be filled with a list of blank space characters, if it were going to hold a list of groups of words. Or it would be filled with a list of zeros (or blank spaces), each prefixed by a single non-alphanumerical character, if it were going to hold a list of Primary Keys.

The number of entries in the list would indicate its maximum length. The number of zeros, in the first entry, would indicate the minimum and maximum values of all the numbers in the list. Similarly, the number of blank space characters, in the first entry, would indicate the maximum length of all the groups of words in the list. And likewise, the number of zeros (or blank spaces) prefixed by a single non-alphanumerical character, in the first entry, would indicate the possible values (or maximum length) of all the *Primary Keys* in the list.

For example, if the first entry in a list had five zeros, then the numbers could range from −99999 to 99999. If the first entry had five blank space characters, then all the groups of words in the list could each be up to five characters long. If the first entry began with a hash, followed by 20 blank spaces, then all the *Primary Keys* in the list could each be up to 20 characters long.

Amongst the hardest set of *data* to hold in a *Database*, based on comma-separated *ASCII* format, would be information that could be used to build a *Graphical User Interface*, such as 2D images, 3D models, *Textures*, points in 2D or 3D space and so on. Since the *ASCII* format was not meant to describe such *Interfaces*, but simple *User Interfaces* based on text. However, there are also several *open data formats*, based on *ASCII* characters, that could be used to hold graphical information. These include the *SVG Format* and the *X PixMap Format*.[36]

There are several CSV *Format sources*[37] that describe how CSV *Formats* could be used to construct a **Game Database**. And *SVG Format sources*[38] that describe how the *SVG Format* could be used to describe graphical information. And *X PixMap Format sources*[39] that describe how the *X PixMap Format* could be used to describe graphical information.

4.1.11 SOUND TOOLS

To record the audio for a game, into a digital *data format*, and manage that *data*, you would use four types of tools.

The first tool would be an electronic device used to record live sounds, onto a suitable media (e.g. *Digital Audio Tape*[40] or DAT or *Secure Digital Card*[41] or SD Card or *Secure Digital High Capacity Card*[42] or SDHC Card). This could be used to either record artificial sounds indoors in a controlled environment, like a sound studio. Or it could be used to record the natural sounds outdoors in an open environment, or an activity, that was going to be portrayed in the Computer Game: such as the roar of a crowd in a stadium. A good example of such a device is the

Zoom H4N Audio Recorder

with a professional external microphone e.g.

Rode NTG4+

and a windshield attachment e.g.

Rode Blimp

to reduce unwanted environmental sounds outside such as the wind.

The second tool would be used to convert these sound recordings, from a media (such as DAT, SD, SDHC or Compact Disc), in a closed or proprietary *data format* (e.g. MP3 or WAV Format) into an open *data format* used for computers (e.g. *Pulse Code Modulation*[43] or PCM). There are *Free software* tools which could do this, such as CDEX and GRIP. There are a few *Digital recording sources*[44] that discuss these tools.

The third tool would be optional. Some computer hardware use a special *data format* for playing back sounds. The hardware vendor would normally provide you with the tool required to convert the sounds, from a standard *data format*, into this proprietary *data format*. This tool normally comes with the proprietary *software library*, from the hardware vendor to the *Software Developer*, to develop games for that proprietary hardware. Once the sound was in the correct *data format*, you would be able to add it to a *Database Record* of the **Game Database**, using the *Database* tools described earlier.

The fourth tool would also be optional. You may either want to listen to each sound, separately or mixed with others, prior to adding it to the *Database*. Or you may want to filter out any unwanted accompanying noises or distort the sound. There are many commercial tools which could be used to mix, alter and listen to such previews. There are also *Free software* tools, such as Audacity (https://www.audacityteam.org/). There are a few *Digital playback sources*[45] that cover these tools.

4.1.12 RUNNING THE TEST

The steps for building the game and the steps for running the test of the game, in the feasibility study, have already been described in the previous subchapters

Step 1: Feasibility Study/Vertical Slice and Designing the Test

If the final step fails, and the game fails to pass the test, then you may need to go back to the previous steps and rebuild the game. To fix the errors reported when the test failed.

How far you would go back in the steps would depend on the errors that were reported. The steps break down into three basic phases which are, in chronological order, building the **Game Database**, building the **Host Modules** and building the **Game Objects**.

If the errors were related to the **Game Objects**, then you would not have to go back that far. If the errors were related to the **Host Modules**, then you would have to go further back. If the errors were related to the **Game Database**, then you would have to go back even further. And after that you would repeat all the steps again forwards leading up to the final step i.e. testing the game.

You may need to repeat this cycle several times before the game successfully passes all the steps of the test without any errors. At which point the time the feasibility study ended should be recorded in the *game design*, alongside the time the study ended.

Now there are some errors you will encounter that you cannot foresee and would be outside of your control: from the third-party software tools, *software libraries*, computer hardware or human error.

But there are some errors or conflicts though which you can foresee and control.

4.1.12.1   Database Host, Graphic and Sound Memory Errors

There could be a potential conflict between how the **Database Host** would be designed, and how some computer hardware display graphics or play sounds. Some hardware store graphic and sound *data* in a separate space, outside the main computer memory, where the rest of the *game* resides. This means that either the **Game Database** would need to be split between the main computer memory, the graphics memory and the sound memory. Or some *Database Records* or *Database Fields* would be duplicated in the main memory, the graphics and the sound memory.

In either case, the **Database Host** would need to intelligently decide when the *Database Records* or *Fields* needed to be moved, or copied, from the main computer memory, to the graphics or the sound memory. It would also need to decide when these *Records* or *Fields* should be transferred back into the main memory or replaced in the graphics or the sound memory.

This problem could be solved using the *Records* in the *Database*, which would indicate the graphic or the sound *data* being used. Remember that the description of the **Graphics Host** and the **Sounds Host** included three *Records*. Two of these would hold a list of 2D or 3D **Game Objects** being displayed. The third would hold a list of sounds being played.

So when any graphic or sound *data* were accessed from the *Database*, the **Database Host** could use these three *Records* to check whether the *data* were in the main computer memory. If the *data* were not, then the **Database Host** could copy the *data* from the graphic or sound memory into a *Database Record* or *Field* in the main memory.

For example, if a **Sound Stream Record** was added to the list of sounds being played, the **Database Host** could automatically copy the *sound stream*, from the *Database Record* in main memory to the sound memory in the Sound Card. And the **Sounds Host** would start playing that *sound stream* on that Sound Card.

Alternatively, when a new entry was added onto one of these three *Records*, the **Database Host** could automatically copy the appropriate *data*, from the *Database Record* or *Field* in main memory to the graphic or the sound memory. And then remove it from the main memory to make space, assuming that there was enough space in the graphic or sound memory to hold that data. And after it had been used, the **Database Host** would copy it back from the graphic or sound memory to a *Database Record* or *Field* in the main memory.

4.1.12.2   Database Host and Limited Memory Space Errors

A similar technique could be used if the **Game Database** proved to be too large to load into the main computer memory. The **Residents List Record** and **Absents List**

Record in the *Database* could be used to keep track of the *Records* which were, and were not, loaded into the main memory.

When a *Database Record* was loaded into the main memory, you could add it to the *Residents List Record*.

When the amount of space taking up by *Database Records* residing in the main computer memory went over some limit, you could remove the oldest unused or least frequently used *Database Record* from the main memory to make space. And add that *Record* to the **Absents List Record**.

When a *Record* was accessed, which was not in the main memory, the **Database Host** would then read it from the media the *Database* was stored on. Just as for the previous problem, the *Database* would need *Records* reserved in it to contain these temporary *data*.

In such cases, most professional *Database* software read blocks of *Records*, rather than individual *Records*. This to limit the number of times the software has to keep accessing the storage media, by assuming that it may soon want to read the neighbouring *Records* as well. But you should only do this too after you have tested your software and found that accessing one *Record* at a time was the main cause of the limits to the performance of the software.

4.1.12.3   Database Host and Restricted Access Errors

To ensure that each *software module*, which used the **Game Database**, got the right *Database Table*, *Record* or *Fields*, there are at least two systems that could be applied. Either each **Host Module** (or **Game Object**) could be carefully designed so that it would only access *Records* whose *data format* it knew. Or the *Database* could be designed so that it had extra information, which identified the types of *Records* stored in each *Table*, and the types of *Fields* in each *Record*.

The first method should be possible if the *Records* were carefully designed and documented. Similar *Records* should begin with the same set of *Fields*, and those of the same type should have exactly the same arrangement of *Fields*. This means that those who wrote the *software modules* could be confident that the *Fields* (i.e. the first, second and third *Field*), within a *Record*, were of a certain type.

The second method would mean adding extra *Records* or **DATABASE META DATA RECORDS** to the *Database*. Each would contain the Primary Keys of all the *Records*, in the *Database*, of a particular type. Each **Meta Data** would also include a list, describing the different *Fields* in these *Records*. The order of the descriptions would match the order of the *Fields*. And each description would indicate whether the corresponding *Field* was a number, a group of words or a Primary Key. Or the description would indicate whether it was a list of numbers, groups of words or Primary Keys.

So when each *software module* accessed the **Game Database**, it would provide the **Database Host** with the **Meta-Data** or type of *Record* it wanted, as well as its Primary Key. The **Database Host** would then use the **Meta Data**, to check that the *Record* was the correct type, before sending it back to the *module*. Once it had the *data*, the *module* could also use the **Meta Data** to find the *data* it wanted. It could use the **Meta Data** to find the position of the *Fields* containing that *data*.

4.1.12.4   Database Host and Corruption Errors

Further additions may be made to the **Database Host** to resolve problems with errors due to corruption of the *data*. Assuming there were no errors in the **Game Database**, when it was originally created, it could be used to detect and correct corruption in the *data* that occurs. Either when running the game in computer memory or when installing the game.

This could be done by simply adding **Database Meta Data Records** or **DATABASE CHECKSUM RECORDS** that contained the *Checksum*[46] for *Database Tables* or *Database Records*. This *Checksum* would be the total value of the contents (i.e. numerical value of the characters or numbers) in the *Database Table* or in each *Record*. You could then detect corruption in the **Game Database**, when either the current total value of the *Database Table* differed from the *Checksum* for the *Table*. Or when the current total value of the *Fields* in a *Record* differed from the *Checksum* for that *Record*. And you could recover from the corruption by simply reading back the *Database Table* or the *Database Record*, from the storage media containing the *Database*.

A *software procedure* could be added to the **Database Host** for achieving this. This could read back one or more *Database Table* or *Records* of the **Game Database**, from the storage media it was stored on. Other *software modules* would either use this *procedure* when these detected corruption in the *data*. Or this may be used when a *module* required the *data* in a *Record* to be reset. So that the *module* could repeat some cycle involving that *Record*. For example, a *module* using one or more *Records* to display the animation of a character, made up of a repeated cycle of *Frames*, may use the *procedure*. To reset those *Records*, when each cycle had been completed.

And when installing the game, you can detect corruption in the **Game Database**, by similarly checking the **Database Checksum Records**. Suppose the *Game World* was divided into several stages, and each stage had a **STAGE OBJECT LIST RECORD** listing all of the other *Records* which should be loaded in memory to play in that stage. And all the **Stage Object List Records** were in a *Database Table* called the **OBJECTS LOADED TABLE**. For examples of **Stage Object List Records** and an **Objects Loaded Table**, please refer to Chapter 1 in the second book in the series, Event-Database Architecture for Computer Games: Game Design and Nature of the Beast, and the subchapter entitled 1.4.19 Objects Loaded Table".

Now after the game is installed, the **Database Host** could automatically check, whether all the *Records* required to play in each stage were present. By checking the *Checksum* for the each of the **Stage Object List Records**. If a *Record* or **Game Object** were missing from the **Stage Object List Record**, then the *Checksum* for that *Record* would reveal this. And if a **Stage Object List Record** were missing from the **Objects Loaded Table**, then the *Checksum* for that *Table* would reveal this. And if a **Game Object** or any of its **Events** were missing, then *Checksum* of the *Database Table* of **Game Object Records** or **Event Records** would reveal this. And any tool, outside of the game, can check the integrity of the **Database**, due to the *open data format* the **Database** would be written in. This includes any tool that either installs the game from a storage media or tests the game before it is put on some storage media and released to the public or tests the game before it goes through some long process of putting it on some *platform* with less resources than the one the game was developed on. However, in contrast, there is no such tool which can check the integrity of the **Game Database** of commercial *game-engines*. There are no such tools that can check the integrity of the **Game Database** after it has been installed from the storage media, before it is put on some storage media, or before it is transferred to another *platform*. There are only tools which check the integrity of the compressed archive, from which the game and the **Game Database** was installed. But that does not mean that the **Game Database** is missing something or is corrupted. That only means the compressed archive is corrupted. If the **Game Database** is missing something, then you will only find out when you play the game, and you reach the point where that missing data was used. And this may appear when it is too late: after a game has been published and released to the public.

4.1.12.5   Events Host, Physics Host and Recursion Errors

The **Events Host** may also be modified to help mitigate errors. These would namely be errors that might occur because of a long, recursive sequence of **Events**. Such a sequence would keep the **Events Host** locked up, indefinitely, repeatedly sending the same cycle of **Events**. However, a limit could be set as to how many **Secondary Events** may be sent in any *Unit of game time*. And if this limit were reached, the **Events Host** could store the remainder. These could be stored in the **Delayed Events List Record**, in the **Game Database**. The stored **Events** would have no time delay and be placed at the front of the list of delayed **Events**. So that, in the next *Unit of game time*, the stored **Events** would be sent first.

Likewise, the **Physics Host** may also be modified to help mitigate errors when updating the physical properties of **Game Object**. If time taken by updating the physical properties of **Objects**, and sending the **Secondary Events** generated from these updates leads to the **Physics Host** exceeding the *Unit of game time*, then the **Physics Host** will stop. It will put all of the **Game Objects** whose physical properties have not been updated onto a **DELAYED 2D PHYSICS LIST RECORD** or a **DELAYED 3D PHYSICS LIST RECORD**. These *Records* will have the same *Database Fields* as the **Physics List Record**. And when the next *Unit of game time* began, the **Physics Host** will resume updating the physical properties of **Game Objects**. Beginning with those **Game Objects** that were placed on one of these two lists. And removing each one in turn from the two lists as its physical properties were updated.

4.1.13   PROGNOSIS FROM THE TEST

After the feasibility study has been completed based on a cross section of the game e.g. 10% of the whole game, it is time to make a prediction or prognosis of how long it would take to build the whole game. Based on the time it took to build and test the small minimal game or cross section of the game, built on the **Event-Database Architecture**. This time should be visible from the times recorded in the *game design* during the steps of the feasibility study as already explained in the previous subchapter **Feasibility Study/Vertical Slice**.

The prediction for the time it would take to build the whole game should be included in the *game design*, in the next step of the **Event-Database Production Process** for the whole game.

This prognosis for the time it would take to build the whole game should be used to make an assessment of whether the overall feasibility study has succeeded or failed. Depending on whether it was possible to execute all of the steps of the test at the end of the feasibility study. And whether this prognosis for producing the whole game produces a time which is within the deadline for the project.

For example, suppose it took three months to build 10% of the game in the feasibility study. The prognosis should then be that it would take 30 months or 2 years and 6 months to make the whole game. If the deadline for the project were two years and six months from now, then overall the feasibility study has succeeded. If the deadline for the project were 18 months from now, then overall the feasibility study has failed.

4.2   STEP 2: GAME DESIGN

The *game design* in the **Event-Database Production Process** may be written in the same way that it would in the normal ad hoc or *Software Evolution Process* used in the Computer Games industry.

This would include a brief outline of the main themes or background story for the game, the main characters in the story and the plot, or the goal in the game.

This would include a brief description of the different parts of the *Game World*, the characters, locations, the animate and inanimate objects in each part. The rules for playing through each part, including the rules of the beginning, middle and end of each part. And the progression of the player from one part to the next.

And this would include a mock-up of the *User Interface* showing the 2D or 3D *User Interfaces*, through each part. This would include rough 2D or 3D artwork sometimes called 'Concept Art' showing how each location of the *Game World* or each menu in the game would look like. And it would include 'Concept Art' showing how some of the characters, locations, animate and inanimate objects in each location or items on the menus would look like.

4.3   STEP 3: TECHNICAL DESIGN

The main objective of the *technical design* is to outline the rules for generating a system of **Events** and the rules for generating a system of **Game Objects** that will be used with the **Event-Database Architecture** to build the game.

To illustrate the application of these rules, the *technical design* may include examples of how popular features of the Computer Games, such as Artificial Intelligence, Physics, Graphics or Testing, could be implemented using these rules.

4.3.1   RULES FOR GENERATING THE SYSTEM OF EVENTS

Any system you choose for adding **Events**, to the **Event-Database Architecture**, would exclude those **Events** required by the **Architecture**. That is to say, it would exclude all of the standard **Events** set out in the description of the **Architecture**. These would be, namely, those listed in the description of the **Events Host** and those that were required by the **Game Controllers Host**.

The latter of these two sets would vary depending on the *User Interface* for the game. Depending on the set of commands that could be issued, by the player, a different set of **Primary** and **Secondary Events** would be required. Remember that any command that could be issued, through the *Game Controllers*, would require one unique **Primary Event** to be sent when it was used. This would also require one unique **Secondary Event**, which would respond to the *analogue devices* or *digital devices* being used, and allow one of the **Game Objects** to recognise that command.

So excluding these standard **Events**, there would be at least three systems for adding new **Events**, you could choose from

1. whenever a *software procedure* (i.e. the **Action** of a **Game Object**) caused a transition in the *data*, stored in one or more *Database Fields*, it should send one unique **Primary Event**, and that *procedure* should only ever send that same **Event**;
2. whenever an **Action** started or ended, that **Action** should send two unique **Primary Events**; one at the beginning and another at the end;
3. whenever a *logic branch* occurred in a *procedure* or **Action**, which was neither part of a check for errors nor part of a search of the *Database*, *two* **Events** should be sent; one for the positive result and one for the negative result of that *branch*; and each **Action** should have only one *logic branch*.

For all three systems, you would initially write all the *software procedures* that a **Game Object** would use to respond to its **Secondary Events**. You would then go back through these *procedures* and insert the steps for sending **Primary Events**, into each one, according to the principles of the system. All three systems would be based on presumptions about where, in a *software procedure*, a change in the flow of the game may occur.

The first system would send a unique **Event** based on the presumption that a change in *data* may cause a change in the flow of the game. The second system would send an **Event** based on the presumption that the start or end of a *software procedure* may cause a change in the flow of the game. And the third system would send an **Event** based on the presumption that a *logic branch* may cause a change in the flow of the game. This change could happen either within the same **Game Object** or within another **Object** of the system.

4.3.2   RULES FOR GENERATING THE SYSTEM OF GAME OBJECTS

The choice of **Game Objects**, used to build a *game design*, would also affect the choices available for making changes to that design. There are at least two systems for adding **Game Objects**, to the **Event-Database Architecture**, you could choose from

1. each feature of a game should have one or more states, and one **Game Object** should respond to each state;
2. each feature of a game should be implemented by one or more formulas, and one **Game Object** should be used to evaluate each formula.

Both systems produce simple **Game Objects** based on presumptions about how the features of a game may be divided into small, predictable software components.

The first system produces predictable **Game Objects** based on the presumption that every feature of a game (i.e. an item lying around the *Game World*, a character, a location, a menu and a picture) may operate in different states. And that the *software module* which implements that feature may be divided into separate **Game Objects,** one for each state. The second system produces predictable **Game Objects** based on the presumption that every feature of a game may be composed from one or more formulas. And one **Game Object** may be used to evaluate each of these predictable formulas.

For the first system, you would begin by writing a single **Game Object** to implement a feature for a game. This **Game Object** would respond to the one or more **Events** which it required to implement that feature. And it would include one or more *software procedures* that would be used to respond to these **Events**. However, instead of using *logic branches* to control the flow of its behaviour, the **Game Object** would be written to operate in different states. Depending on its current state, each *software procedure* would behave one way or another. And this state would be stored in one of the *Fields*, of the *Record* which held the properties of the **Game Object**. The **Object** would check, in its *software procedures*, when it needed to change to another state, and it would change state when necessary.

After writing this **Game Object**, you would replace it with one or more new **Game Objects**. You would begin by examining how many different states the original **Game Object** used. And you would write a new **Game Object** for each state. Each new **Game Object** would have the same properties as the original one, excluding the *Field* indicating its state. Each would also respond to a unique set of **Primary Events**, similar to those of the original **Game Object**. However, each new **Game Object** would only respond to those **Events** required for the state it represented.

Likewise, to respond to these **Events**, each new **Game Object** would use similar *software procedures* to the original **Object**. Except that there would be two differences. The first would be that each **Object** would only use the *software procedures*, or parts of the *procedures*, used within its state. The second difference would be that, when it was time to change to another state, each **Game Object** would simply send one unique **Primary Event** for the new state, instead of changing a *Field* in the **Game Database**.

When this happened, the old **Game Object** would respond to this **Event** and disappear from the *Game World*. At the same time, the **Game Object** of the next state would appear, in the same location as the old one. Or, if in fact these **Game Objects** were never visible in the *Game World*, the old **Game Object** would simply stop responding to the **Primary Events** that it shared with all the others. And the **Game Object** of the next state would start responding to these **Events** in its place.

4.3.3 APPLICATION: TESTING

Whatever system of **Events** or **Game Objects** you choose, these should be used consistently to add all the features of a *game design*. These systems should be used regardless of whether the person writing the **Game Objects** thinks others may, or may not, find those **Game Objects** or **Events** useful. That decision should be left to the *Game Producers* and *Designers*, who would refer to the **Game Database** to see what **Game Objects** and **Events** were available. And would modify the *Database* to achieve the combination of **Game Objects**, or the sequence of **Events**, they wanted.

Besides providing this basic ability to easily modify the game, when the *game design* changed, there would be three more advantages to using these systems consistently. Firstly, it would simplify the process of translating the features, from a *game design*, into the *technical design*. Secondly, it would prevent conflicts that may arise by people using different systems for adding **Events**. And finally, it would help debug errors in a feature of a game, since you could use these systems to deduce what kind of **Events** or **Game Objects** were used to implement it.

The first set of **Events** and **Game Objects**, of the production process, would be enough to cover the initial *game design*. Unless this *game design* was completely wrong, it would also be possible to re-use some of these **Events** and **Objects** to partially implement any change. So most new **Objects** should try to re-use **Events** and **Actions** from existing **Game Objects**. This is the reason for the inclusion of step (14) in the **Event-Database Production Process**, which would identify the existing **Objects** that could be re-used.

It would be useful to document each initial feature, or change, in a *game design* with a flow diagram. The diagram could list the *Records* and **Game Objects** that were used. And it could show the **Events** which connected the **Actions** of the **Game Objects** together. It would also be useful for the *Database Administrator* to document all the new *Records*, in the *data design*, for future reference. This would preferably always be done before these were added to the **Game Database**, during the design of the *Records*, in step (3) or step (16) in the **Event-Database Production Process**.

The extent of the range of documentation available would be the difference between the process for producing games, used by the **Event-Database Architecture**, and a completely formal production process. In a formal process, such as the *classic software production life cycle*, when changes need to be made to a game, you would have to revert back to step (1). This would entail modifying the high-level *game design*, the low-level *technical design*, and perhaps the *data design* and the *tools design*. Finally, you would build the new components required to implement the changes, test each new component separately, before combining these together.

However, in the **Event-Database Architecture**, you would not need to revert back to step (1). You would only need to update the *data design*, implement and test each new component, before using these to change the game. The reason for this has already been mentioned when the initial problem, the **Event-Database Architecture** was meant to solve, was described. Once the *data* was well-defined, it would be easy to deduce how any software which used these functioned or should function. This would only require a general knowledge of the purpose of that software.

Nevertheless, if you preferred, you could keep all the design documents up-to-date. A formal production process, such as the *classic software production life cycle*, would still benefit from the **Architecture**.

An important part of the *classic software production life cycle* would be the testing of the software at the end of it. This would involve testing every possible sequence of actions, or decisions, that could be made in the software. You would have to traverse every path, from the root to a leaf, in the tree of possible sequences. The root would be when the software was started. Each branch would be when a decision was made about the next sequence of actions, mainly by a *logic branch*. And each leaf would be when you reached a point where you had no option, but to return to a previous stage, or node in the tree of possible sequences of actions or decisions. The **Event-Database Architecture** may allow you to test every one of these *logic paths*,[47] depending on two choices you could make.

Firstly, you could choose a system for adding **Events**, to the **Architecture**, that revolved around *logic branches*. So that each *branch* was performed by using a **Primary** and a **Secondary Event**. As a result, each **Event** would represent a branch in the tree of all the possible sequence of actions or decisions that could occur in the software.

Secondly, you could choose to comprehensively document the interconnection between these **Events**, so that you could understand the whole tree. Nevertheless, even if the documentation did not permit this, the **Architecture** would still allow you to test every branch in the tree. Since each **Event**, in the **Architecture**, would represent a branch in the tree, you could test each one separately.

Documenting all the interconnections between **Events** would be, of course, the better of the two options. It would allow you to perform a more thorough and exhaustive test. However, another alternative to producing these documents would be to use the **Game Database**. There would be at least two ways in which the *Database* could be used to track the interconnection of **Events**.

One way would be to use the **Log Records** of the **Database Host**. These could be used to monitor the *Records* of **Primary Events**. This would keep a log of all the changes to the properties of **Primary Events**. So, assuming that each time a **Primary Event** was used, its properties were changed, a log would be kept. And so, you could use these logs to trace the chain of **Events** that occurred during a game.

But this would require every *software module*, in the **Architecture**, to modify the *Record* of a **Primary Event** before using that **Event**. As it happens, the software components of the **Event-Database Architecture** would do just that. The **Physics Host**, the **Sounds Host** and the **Game Controllers Host** would all modify the *Record* of a **Primary Event**, before sending that **Event**. And the list of **Secondary Events** these three added, to the properties of the **Primary Event**, would be those that were about to be used.

However, there would be no requirement for the **Game Objects** to do this as well. Unless, that is, a new rule were added to the system of generating **Game Objects**. To compliment the system of generating Events. This rule would be that every **Game Object** should get the **Secondary Events** that it would send, from specified *Records* in the **Game Database**. And it should only add these **Events**, to the properties of a **Primary Event**, when it wanted to send that **Primary Event**. After it had finished using that **Primary Event**, the **Game Object** should remove all the **Secondary Events** that it added.

The advantage of using the **Log Records** of the **Database Host**, to find the chain of **Events** used for a game, would be that the method would be automated. By simply playing through a game, you could collate its chain of **Events**. The disadvantage would be that how much of the overall chain was revealed would be limited. It would be limited to the areas of the game you played through.

Another way of getting this chain would be to add two more properties, to the *Record* of each **Primary Event**. The first would be a list of all the **Game Objects** which send that **Primary Event**. The second would be a list of all the possible **Secondary Events**, which were going to be sent in response to that **Primary Event**, during a game.

Using these two new properties, along with those of **Primary** and **Secondary Events** already in the **Game Database**, you could trace the lineage of every **Event**. For every **Secondary Event**, you would be able to see which **Primary Events** used it, from the *Database*. For every **Primary Event**, you would be able to see the set of **Game Objects** which sent it. And for each **Game Object**, you would be able to see the set of **Secondary Events** it would receive. This last information would come from the **Secondary Events Record** (see Section 3.1). So, for any **Event** in the middle or end of the game, you would be able to trace back the chain of **Events** from the **Primary Initial Reset Event**, which started the entire game to that Event. You would be able to collate all the sets of **Primary Events**, **Secondary Events** and **Game Objects** involved in the chain.

This method would have the same advantage as using the **Log Records** of the **Database Host**. Namely, it would allow you to automatically collate the chain of **Events** used in a game. It would have the added advantage in that it would not be limited, to which areas of the game you have played through. It would collate the chain of **Events** in all areas of the game.

For the debug version of this game, another advantage of this method would be that it could help prevent errors. Each **Primary Event** would have a list of the **Game Objects** that would send it. So you could stop the misuse of a **Primary Event**, by **Game Objects** which were not on that list. Each **Event** would also have a list of **Secondary Events** that were going to be associated with it. So you could ignore any **Secondary Event**, associated with a **Primary Event**, which was not on that list.

The disadvantage of this method, however, would be its accuracy. The chain of **Events**, which it would construct for a game, would actually be a set of possible chain of **Events**, from the **Primary Initial Reset Event** to any other target **Event**. But only a subset of this will actually be used in the game. So you would have to try out the different possibilities and eliminate the ones which did not lead to that other target **Event**.

At each point along the chain of **Events**, from the start of the game, there would either be a set of **Game Objects**, a set of **Primary Events** or a set of **Secondary Events**. Each member of a set of **Primary Events** may produce one or more of the next set of **Secondary Events** in the chain. And one or more of the **Secondary Events**, within a set, may be received by one member, of the next set of **Game Objects**, in the chain. It would be possible to tell which **Game Object** responded to any given **Secondary Event**. But it would not be possible to tell which **Primary Events** were subsequently produced, by that **Game Object**.

So to find the exact path, from the start of the game, to any other **Event** you were interested in, you may have to experiment. For any given set of **Secondary Events**, you may have to experiment with the members of that set. It may take several attempts to find the **Secondary Events** which produced one of the next set of **Primary Events**, in the chain of **Events**. A **Game Object** may either only produce a **Primary Event** after a **Secondary Event** has been received several times. Or it may only produce a **Primary Event** when some previous **Event** had occurred several times. Or it may only produce a **Primary Event** when some combination of previous **Events** had occurred.

Therefore, beginning with the **Primary Initial Reset Event**, you would have to experiment with all of the first set of **Secondary Events**. You would have to select the ones which, when received by its **Game Object**, produced one of the next set of **Primary Events** in the chain. And you would eliminate those **Secondary Events**, from the first set, which did not produce any of the expected **Primary Events**.

Similarly, for each **Event**, in the first set of **Primary Events**, you would have to experiment again. You would have to select its **Secondary Events** which produced one of the next set of **Primary Events**, in the chain. But this time, if none of its **Secondary Events** did this, that **Primary Event** would have to be eliminated from its set. You would repeat this process for all the subsequent sets of **Primary Events**, along the chain of **Events**.

If after eliminating a **Primary Event** or **Secondary Event** from its set, that set were to become empty, this would mean that you had reached a dead end. This would be either because one of the previous **Secondary Events** along the chain had to be repeated several times. Or this would be because some combination of these previous **Events** had to occur.

So you would have to repeat the entire search through the possible chain of **Events**, from the start of the game. But, this time, you would either have to increase the repetition of some **Secondary Events**. Or you may have to try some of the **Secondary Events** in a different order. Hopefully, the description of the *Records* of the **Primary Events** and **Secondary Events**, in the *data design*, should help you pinpoint exactly which **Events** had to be repeated, or the order of **Events** required, to unlock the next level down in the chain of **Events**.

The search would stop once you reached the end of the chain of **Events**. At which point all the possible paths, from the start of the game to end of the chain, would be any combination of the remaining **Primary** and **Secondary Events** in the sets, along the chain.

The entire search for these paths may be partially, or completely, automated by using a tool. This tool could be specifically built to help with the testing of a game, at the end of a *software production life cycle*. One, or more **Game Objects**, could be added to the **Event-Database Architecture**, to build this tool. And it would be possible, for a player, to access this tool at any stage of the game. So that either the player could use the tool manually, to test the effect of any **Primary** or **Secondary Event**, at that stage. And thus the player could find a path, from the start of the game, to any other **Event**. Or the player could manually describe to the tool, the sets of **Primary Events**, **Secondary Events** and **Game Objects**, from the start of the game, to any given **Event**. And the tool would systematically search through all the possible chain of **Events** and select all the feasible ones.

The time it would take to find the feasible paths would depend on how the **Game Objects** were built. It would depend on the average number of **Secondary Events** each **Game Object** would receive. And it would depend on the average number of **Primary Events** each **Game Object** would generate. The higher these numbers were, the longer time it would take to find which **Secondary Events** produced which **Primary Events**. The lower these numbers were, the less time it would take to get this information. And ultimately, for optimum time, it would take to perform this search, each **Game Object** should receive only one **Secondary Event** and send only one **Primary Event**.

4.3.4 APPLICATION: GAME PLAY – ESCAPING A PRISON

Compare how the three systems for generating **Events** mentioned in the subchapter entitled **Rules for Generating the System of Events** would be used in a production process, based on the **Event-Database Architecture**, with the *Software Evolution Process.*

For example, suppose there were one stage of a game in which the player would be imprisoned, and held for ransom, in a cabin, on board a pirate's ship. But in that cabin there would be three open chests, each full of treasure taken by the pirates. And, in order to escape, the player had to throw all of this treasure overboard, through a port hole left open in the cabin. The contents of each chest would be so full that it would take several attempts to empty it all. After emptying all of the chests, and leaving two of these open, the player had to climb into the remaining chest and close it. So that when the captors entered the cabin, they would panic, believe the player had escaped and start frantically searching the ship, leaving the door of the cabin open. Thus, they would present the player with an opportunity to escape.

Now in the *Software Evolution Process*, this stage of the game would be built over several steps and evolve each time. The first step would be to add the cabin to the *Game World* and display it. This would include the background scenery around the cabin, such as the rest of the ship, the sky and the sea. The second step would be to place the three chests into the cabin. The third would be to introduce the player's character. This would include the commands that would move this character around the cabin. And this would include other features such as the animations and sounds, which would be seen and heard when the player walked around.

The fourth step would be to add the commands that would open and close the chest, along with the animation and sounds for these actions. The fifth would be to add the different quantities (i.e. images) of the treasure in each chest, from an amount filling up half of the chest to an amount that was virtually overflowing. The sixth step would be to add the command that would partially throw out the contents of each chest. The seventh would be to add the commands that would allow the player to climb in and out of each chest. Finally, at the end, the animations and sounds that would be played back, when the captors entered the cabin, would be added and displayed in the *Game World*.

But, as would be typical of the *Software Evolution Process*, each of these eight steps would be conducted in an ad hoc fashion. And there would be little or no regard for the next, causing at least three immediate problems.

The first of these problems that would arise would relate to the tools of the *Software Evolution Process*.

Take, for example, the progression from the first to the second of the eight steps just mentioned. After the cabin and its surrounding scenery had been built by the

Game Artists and displayed in the Game World, the three chests would be added with little regard for the relative scale of these to the cabin. In the worst case, one of the Game Programmers would be expected to adjust the size of the chests, by editing the software and rebuilding it. This may take several attempts before the Game Artists, the Game Designers, the Game Producer and anyone else with artistic leanings would be satisfied. In the best case, one of the Game Programmers would be asked to write a new tool. But only after the second step had begun, and it had become self-evident what a waste of time it had been for the Programmers to have to keep editing the software. This new tool would permit the Game Artists, or anyone else, to set the relative scales of items displayed in the Game World and preview these side by side.

Nevertheless, even in this case, the Programmer who wrote the tool would typically also be the one who would write the game module, which used the data produced by the tool. And this data would not be documented. So only that Programmer would understand the tool, the data and that game module. In other cases, you may get the marginally better situation where two Programmers would share this knowledge. Namely, one Programmer would write the tool and another would write the game module.

But, in almost all cases, there would still be no documentation of the tool, or the game module. And the other Programmers, Game Artists or Game Designers would be expected to familiarise themselves with either of the two Programmers. Or with the tool by playing around with it, specifically its User Interface. Through this casual acquaintance, they would be expected to speculate about the design of the tool, the data or the game module. Thus, the User Interface of the tool in effect would become a form of de facto documentation. The various esoteric names, given to the components of this Interface, would literally become components of the language used, by the staff, to communicate. This would contribute to the degeneration of that language, and the overall language of the staff as a whole. And the more this language degenerated, the less productive the Game Programmers, Game Artists, Game Designers and others would become.

The second problem that would immediately arise, from the ad hoc fashion in which the Software Evolution Process progresses, would be reflected in how the flow of the game developed. The process would affect this development in four ways.

Firstly, it would turn the flow of the game from an abstract concept, into a physical one. That is, the flow of any game should be an abstract concept. It should be an intangible high-level component composed of other, more simpler and tangible, low-level ones. And intuitively, you would expect it to be controlled by a high-level software module, which was composed of other low-level ones. Yet counterintuitively, in the Software Evolution Process, the flow of the game would end up being controlled by the low-level software modules: not the high-level ones. That is to say, it would be controlled by modules which had a physical component: either visible, audible or tactile in the game. Since, in the Software Evolution Process, these modules would be the first ones built by the staff. So that the Game Designers and Producers could have some feedback from the software as quickly as possible. And they could start filling in the huge gaps in the incomplete game design the process would begin with.

These physical components, however, would become the basis for all subsequent additions, including the flow of the game.

This leads onto the second effect that the *Software Evolution Process* would have on this flow. That is, the flow of the game would not be directed by one, or a small set of high-level *modules*, with no physical component. Instead, the flow would end up being spread around a large set of low-level *modules*, controlling items in the *Game World*. Thus making the process of editing the flow, far more complex than it needs to be, due to the larger number of low-level *modules* it depends on. Thus also making editing far more risky than it needs to be. Since editing one of these *modules* could produce unwanted side effects on either the flow of the game or the physical appearance of the *Game World*.

The third effect on the flow would be that the set of low-level *modules* directing it would be arbitrary. These *modules* would be the first set which happened to be written to demonstrate the *Game World*. And the flow of the game in all the higher-level *modules* written after them would end up revolving around this arbitrary set. The internal flow of the game, within these low-level *modules*, would end up dictating the external flow of the game around them, in the higher-level *modules*. The structure of the internal *data*, within these low-level *modules*, would end up dictating the external structure of *data* for subsequent higher-level *modules* written after them. And without these low-level *modules* the flow of the entire game would grind to a halt.

The final effect of the *Software Evolution Process*, on the flow of the game, would be that these preliminary *modules*, which are low-level *modules*, would not be documented. Thus, as has already been mentioned, the authors of subsequent higher-level *modules* would waste time duplicating the testing for each branch in the flow of the game, within this first set. Since they would be ignorant of which and how many branches were already detected by these low-level *modules* or other higher-level *modules*.

Take the earlier example based on pirates. It would be counterintuitive to expect either the sky, the sea, the pirate's ship, their prisoner, the cabin or one of chests in it, to control when that stage of the game began. Nor would one of these items be expected to control how long the stage would last, how much time the player had to complete it, before moving on to the next. Nor would these be expected to control when the prisoner had completed the first phase of the escape and emptied all the chests. So that the second phase could begin and captors could come in.

But this would be exactly what would happen in the *Software Evolution Process*! One or more of the *modules* (or **Game Objects**), which controlled the appearance of these items, would be edited to control the flow of the game: from one phase to the next: from one stage to the next. That is to say, you will find either the **Game Object** of the sky, the **Game Object** of the sea, the **Game Object** of the ship, the **Game Object** of the prisoner, the **Game Object** of the cabin or the **Game Object** of the treasure chests controlled the flow of the game in this phase. Precisely which one controlled this phase of the flow would be arbitrary, dependent on the order in which these *modules* were built during the process.

The third problem that would immediately arise from the ad hoc fashion in which this stage of the game would be built, in the *Software Evolution Process*, would be evident in how the *software procedures* were written. These too would be constructed with little regard for the successive steps of the production process.

For example, one way in which the player's escape, from the cabin, would be monitored would be through a single *software procedure*. This *procedure* would be used either periodically, after the stage had begun, or each time the player had closed one of the chests. So that the *procedure* would then inspect the *Game data*. It would check whether the player was in the chest, whether the other two chests were open and whether all three were empty. But the use of a single *procedure* for this purpose would prove problematic later on. For as the *game design* changed, the criteria defining a successful escape would become more complicated. And this same single *procedure* would be edited again and again. It would keep growing larger and larger, becoming more and more complex and unmanageable in the process, with each change in the criteria for a successful escape.

In contrast, with the **Event-Database Architecture**, these three problems would be mitigated. Each step of the production process would be directed by the **Architecture**.

Firstly, the *data* produced by any tool, and subsequently used by a **Game Object**, would be documented in the *data design*. This would include any new tool, written for the *Game Artists*, to allow them to change the scale of the cabin or the chests, and preview the results. If indeed such a tool were necessary, in this example. Since the *open data format*, of the **Game Database**, would mean you could use any *Database* software to edit such information. With the Event-Database Architecture, you can use tools, simpler than programming tools, to query and edit the Game Database. Without having to wait for programming tools to build the game (which in the case of modern commercial game-engines can take between 18 minutes to 14 hours for a game with 300 staff). And then examine the results to see whether these were what you expected. And then edit the Game Database again. And then build the game again. And then examine the results again, and so on and so on. Instead, you could build a more simpler tool Not necessary because of preceding amendment. You could build one that would merely preview any changes which had been made, by reading the relevant *Records* from the *Database*.

Secondly, in the **Architecture**, the beginning, middle and end of the flow of the game would be controlled by one simple component. This component would always be the same, and it would be an abstract, high-level component. It would have no physical component in the game. This component would not be arbitrary and it would be the same in every game based on the **Architecture**. This component would also save time searching through software modules for previous tests for branches in the flow of the game which you want to reuse in a new software module, and eliminate the duplication of such tests. This component would be the **Game Database** of the **Architecture**.

Thirdly, under the first system for adding **Events**, to the **Architecture**, you would never use a single *software procedure* for monitoring the player's escape, as you would in the *Software Evolution Process*. Instead, you would have multiple *procedures* or Actions, each responding to the transition of *data* during the previous step of the escape. And each would send one unique **Primary Event** when the transition for the next step had occurred.

That is, excluding the standard **Events** and **Actions** required by the **Architecture**, this example would require five additional **Primary Events**, with four subsequent **Secondary Events**. And each of these would require four corresponding **Actions** (i.e. *software procedures*) to respond to these **Events**.

You would require one Action that would respond to the contents of a chest being partially thrown out, by the player. And this would send an **Event** when that chest had been completely emptied, as well as increasing the total that had been emptied. You would require another one, which would respond to a chest being completely emptied, or one of the chests being closed. And this would send an **Event** when all three chests had been emptied. You would require another one that would respond to all three chests being emptied and send an **Event** when it detected the player was inside one of the chests. Finally, you would require another Action that would respond to the player climbing into and closing an empty chest, after all three had been emptied. And this would also bring the captors rushing into the cabin to begin their search.

In other words, you would require one additional **Primary Event**, which would be sent by the standard **Action**, required to respond to the command that partially emptied the contents of a chest. This, in turn, would require a subsequent **Secondary Event** with a corresponding **Action**, which would respond after the contents had been partially emptied. This **Action** would send another additional **Primary Event**, but only when that chest had been completely emptied. And it would increase the total that had been emptied, prior to sending this **Event**.

Likewise, this second **Primary Event** would require a subsequent **Secondary Event** and **Action**, which would respond to one of the chests being completely emptied. This **Action** would send a third additional **Primary Event**, but only when all three chests had been emptied.

This third **Primary Event** would also require a subsequent **Secondary Event** and **Action**, which would respond after all three chests had been emptied. This **Action** would, in turn, send a fourth additional **Primary Event**, but only when it detected the player had climbed inside one of the chests.

This would be the final **Primary Event** before the player completed initial phase of the escape. Like the others, it would require a subsequent **Secondary Event** and **Action**, which would respond after the player had climbed into one of the chests. Although this would be the final one, before the completion of the escape, it would not be the final additional **Primary Event** required to build this stage of the game.

Clearly, the chain of **Events**, following on from the command to empty one of the chests, should not lead to a successful escape. For a successful escape, the player would have to climb into one of the chests, after emptying all three, and close the lid. So the chain of **Events** leading to a successful escape should follow on from the command to close the chest. And, hence, a fifth additional **Primary Event** would be required.

This **Event** would be sent by the standard **Action**, required to respond to the command that closed one of the chests. And the aforementioned **Secondary Event**, which would be sent after the player had emptied one of the chests, would also follow on from this same **Primary Event**. This would result in a chain of **Events** which would begin with the player closing the chest, followed by a check of whether all three chests had been emptied, followed by another check of whether the player had climbed into one of the chests and ending with the final **Primary** and **Secondary Events** being sent.

There is an example of the **Primary Events**, **Secondary Events** and **Actions** required to implement this phase of the game in Figures 4.13 and 4.14.

The last of these **Events** would signal that it was time for the final phase of the escape. That is to say, it was time for the captors to enter, find the cabin empty and start a frantic search, leaving the doors open for the player to escape. Of course, this phase could simply be played out, by the last **Action**, through some pre-recorded computer animation. Alternatively, thanks to the **Event-Database Architecture**, it could just as easily be played out interactively; by following the same principle used to build the initial phase of the escape.

That is to say, you could have the **Action** which responded to the chest being closed, sending an **Event** when the captors had entered, after a brief interval had passed. You could add another **Action** which responded to the captors entering

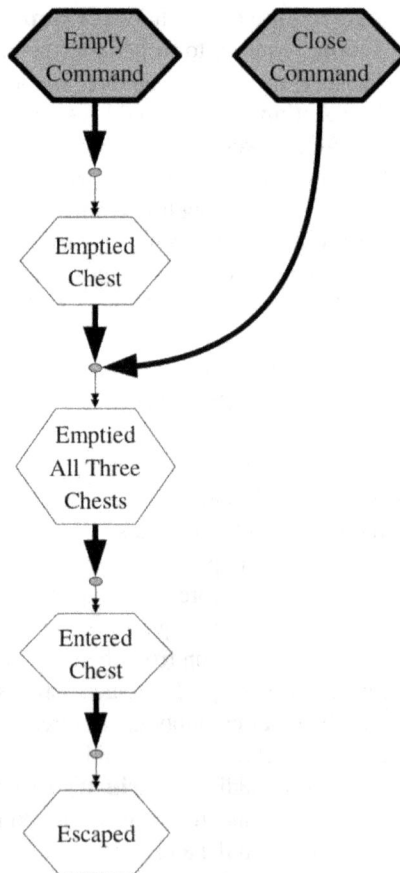

FIGURE 4.13 An example of Primary Events, Secondary Events and Actions that could be generated using a simple rule or system. That in turn could be used to build one quest of a game. This quest being trying to escape from captivity in a cabin, with three chests full of treasure, on board a pirate's ship.

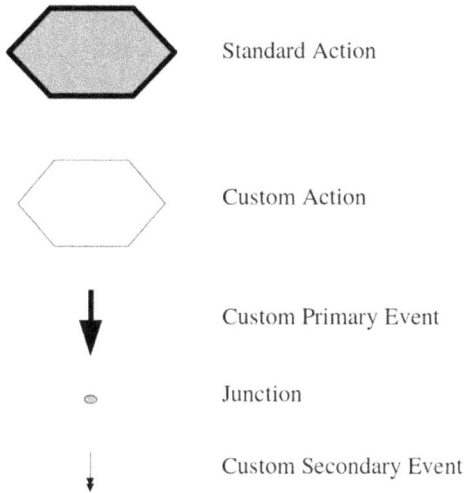

Standard Action

Custom Action

Custom Primary Event

Junction

Custom Secondary Event

FIGURE 4.14 Legend of symbols displayed in Figure 4.13. It is a list of the symbols, for the components of the Event-Database Architecture, that could be used to build one stage of a game.

the cabin and sending an **Event** when they had left. You could add another **Action** which responded to their departure and sending an **Event**, when the player had stepped back out into the cabin and walked up to the open door. And finally, you could add another **Action** which responded to the player stepping out of the cabin, through that door.

Or in other words, the last **Action** which responded to the chest being closed, by the player, could send an additional **Primary Event**, with a subsequent delayed **Secondary Event**, for the captors to enter the cabin. After a brief interval had passed, you could add another **Action**, which responded to this delayed **Event**, and played back the animation of the captors rushing in. At the end of that animation, the standard **Primary End Event** should be sent.

Remember that if, for example, the **Graphics Host** were responsible for playing back an animation sequence, then it would behave in a similar fashion to the **Sounds Host**. That is, once it had finished playing back any pre-recorded sequence, it would notify any **Game Objects** that may want to follow on from that sequence. It would append the **Secondary End Event**, assigned to that sequence, onto the list of those following the **Primary End Event**. And it would then send that **Event**.

So another additional **Secondary Event** could be included, amongst the properties of the animation, of the captors rushing in. This would act as its **Secondary End Event**. This would require a corresponding **Action**, which would respond to the captors leaving the cabin. And this penultimate **Action** would decide what would happen once the player had reached the open door.

It would require another additional **Secondary Event**, which would be received, by one of the **Game Objects**, when the player reached the door. And it would either require that whatever **Action** existed, for responding to the **Secondary Proximity Event** of the door, send an additional **Primary Event**. So that it could append its **Secondary Event** onto the list of those following that **Event**. Or, if no such **Secondary Proximity Event** existed, then it would merely require that its **Secondary Event** be used instead. This **Secondary Event** would, in turn, require the final **Action**, which would respond to the player reaching the door and escaping from the cabin. So in total, including the initial phase of the escape, you would require six or seven additional **Primary Events**, with seven or eight **Secondary Events** and corresponding **Actions**.

In Figures 4.15, 4.16 and 4.17, there is an example of how you would implement this phase of the game, without using computer animation to automate the second

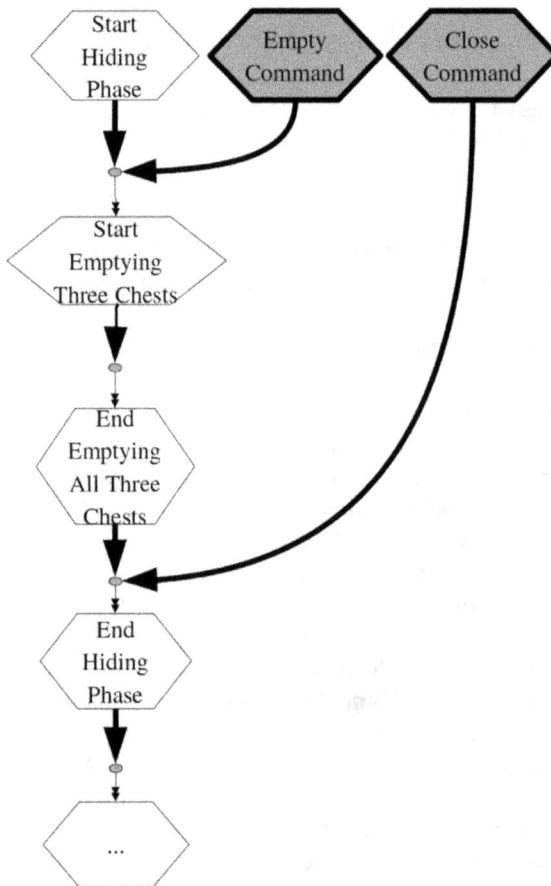

FIGURE 4.15 An example of Primary Events, Secondary Events and Actions that could be generated using a simple rule or system. That in turn could be used to build one quest of a game. This quest being trying to escape from captivity in a cabin, with three chests full of treasure, on board a pirate's ship.

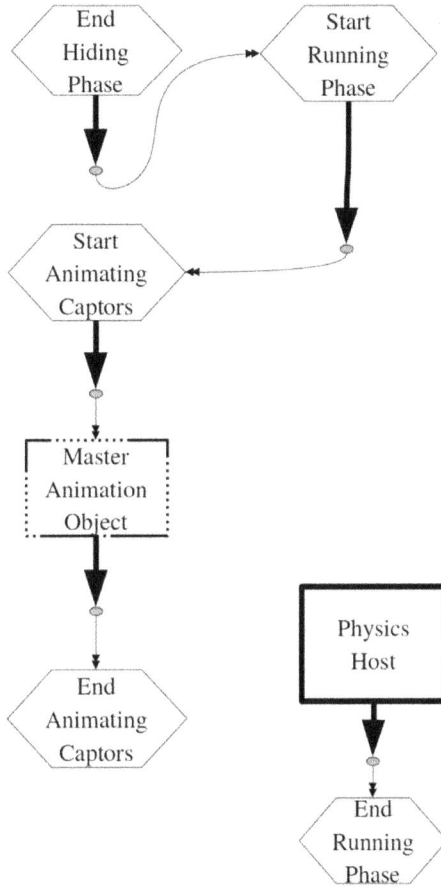

FIGURE 4.16 Extension of Figure 4.15.

part of the escape. And instead require the player to manually complete the second part of the escape (Figure 4.17).

All of these **Events** and **Actions** would leave room for interesting twists that could be made to the *game design* by the *Game Designers*, or others, later on. For example, it could be decided that the player may be unlucky. And that sometimes, when the chests were being emptied, the captors would notice the cargo being offloaded and become alarmed. They would then barge into the cabin, proceed to manhandle the player's character and knock that character out.

This change could be partially achieved by simply editing one of the **Game Objects**, and adding a new Action. This Action would respond to the existing **Events** sent after the contents, of one of the chests, had been partially emptied. It would do little, except just occasionally send another existing **Event**. Namely, it would send the **Event** for when the player had entered, and closed one of the empty chests, having completed the initial phase of the escape.

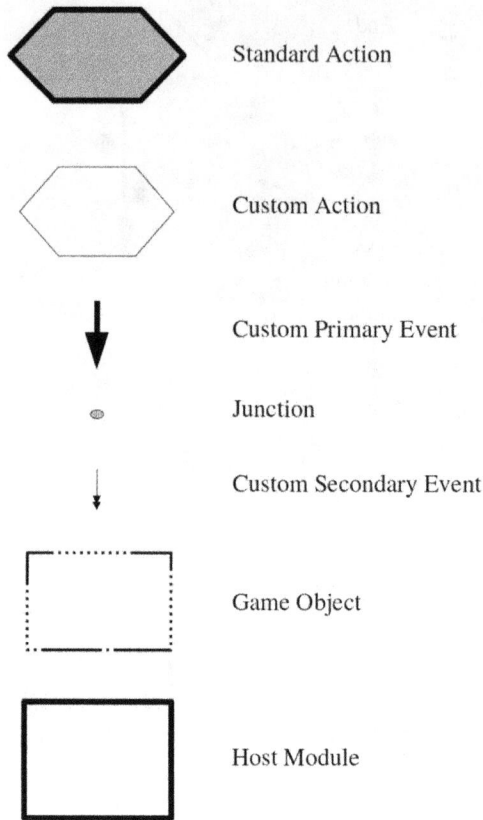

FIGURE 4.17 Legend of symbols displayed in Figure 4.15. It is a list of the symbols, for the components of the Event-Database Architecture, that could be used to build one stage of a game.

In other words, you could add a new **Action**, to one of the **Game Objects**, which would respond to one of the existing **Primary Events**. Specifically, it would respond to the one sent by the standard **Action**, which performed the command that partially emptied each chest. This new **Action** would require a new **Secondary Event**, which it would respond to, that would be added onto the list of those following on from that **Primary Event**. And it would require a new **Primary Event**, which it could send, that would have just one existing **Secondary Event**. Namely, the one that would normally only be sent when the player had completed the initial phase of the escape. So that, in response to the new **Secondary Event**, this new **Action** would occasionally send the new **Primary Event**.

The effect of this change would be that, whenever the player threw out the contents of a chest, the captors would occasionally enter the cabin. And they would act as if the player had escaped. Except this time, it would only be part of their treasure which had escaped.

Of course, they would subsequently walk out and leave the door open, just as if the player had escaped. So to finish off the effect, new animation would be required, which would follow on from the existing animation. And an additional **Secondary Event** would also be required. This would be sent after the existing animation, of the captors entering the cabin, had finished. And it would be appended onto the list of those following the **Primary End Event**, which should mark the end of that animation. The existing **Action**, which played back that animation, could be edited for this very purpose. Furthermore, an additional **Action** would be required that would respond to this new **Secondary Event**. And it would play back new animation, of the captors reentering the cabin, proceeding to manhandle the player and knocking the character out. Thus in total, including the initial and final phases of the escape, you would require seven or eight additional **Primary Events**, with nine or ten **Secondary Events** and corresponding **Actions**.

Under the second system for adding **Events**, to the **Event-Database Architecture**, you would require less **Actions** than the first, to build such an example. You would merely require eight **Actions**. And each one would respond to a unique pair of **Primary** and **Secondary Events**.

You would require one *procedure*, or **Action**, for responding to the **Events** sent after each chest had been partially emptied. You would require another, to respond to the **Events** sent when the captors had noticed the cargo being offloaded by the player. And you would require another, to respond to the **Events** sent after the captors had left and then returned into the cabin, to manhandle the player. You would require one **Action**, for responding to the **Events** sent after the player had entered one of the chests. And you would require yet another, to respond to the **Events** sent after the player had climbed back out of one. You would require one **Action**, to respond to the **Events** sent after one of the chests had been closed. And you would require another, for responding to the **Events** sent when the initial phase of the escape had been completed. Finally, you would require one **Action**, to respond to the **Events** sent when the final phase had been completed.

In other words, under the second system, every **Action** performed in response to one of the standard **Events**, required by the **Architecture**, would also be required to send two unique **Primary Events**. Since every **Action**, would be required to send one **Primary Event** when it began, and another when it ended.

So excluding the standard **Events** and **Actions**, to build this stage of the game, you would require one additional **Secondary Event**. This **Event** would follow on from the additional **Primary Event**, sent at the end of one of the standard **Actions**. Namely, the one that responded to the command for partially emptying a chest. And this **Secondary Event** would require a corresponding **Action**, which would count how many chests had been completely emptied.

You would require another additional **Secondary Event**, which would follow on from the same **Primary Event** as the first one. And it too would require a corresponding **Action**. This **Action** would decide whether the player had only partially emptied the contents of a chest. And it would occasionally alert the captors to the cargo being thrown overboard, if that turned out to be the case. At which point, it would play back the animation of the captors entering the cabin to begin a frantic search. As under the first system, a **Primary End Event** would be sent at the end of this animation.

So a third additional **Secondary Event** would be required to act as the **Secondary End Event** for the animation. This **Event** would require a corresponding **Action**. This **Action** would, once the captors had left the cabin to begin their search, play back the animation of the captors reentering to manhandle the player.

You would require a fourth additional **Secondary Event**, which would follow on from another additional **Primary Event**. This **Secondary Event** would, eventually, follow on from the **Primary Event**, sent at the end of one of the standard **Actions**. Namely, the one performed in response to the command for climbing into a chest. And this **Secondary Event** would require a corresponding **Action**, which would prepare the command for closing the chest, to begin the final phase of the escape.

But, initially, that **Secondary Event** would not be on the list of those following on from the command for climbing into a chest. Until, that is, all three chests had been emptied. At which point, the aforementioned **Action**, which followed on from the command that emptied the last chest, would add that **Event** onto the list.

You would require a fifth additional **Secondary Event**, which would follow on from another additional **Primary Event**. This one would follow on from the **Primary Event** sent at the end of another standard **Action**. Namely, the one performed in response to the command for climbing back out of a chest. And this **Secondary Event** would require a corresponding **Action**, which would stop the closing of the chest beginning the final phase of the escape.

This would be done by removing a sixth additional **Secondary Event**, which would be required, from the list of those following on from yet another additional **Primary Event**. This sixth one would, eventually, follow on from the **Primary Event** sent at the end of another standard **Action**. That is, the one performed in response to the command for closing a chest. And this **Secondary Event**, in turn, would require a corresponding **Action**, which would respond to the end of the initial phase of the escape. This **Action** would also begin the final phase, by playing back the animation of the captors, entering the cabin to begin a frantic search, after the player had hidden in one of the chests.

But, initially, this **Secondary Event** would not be on the list of those following on from the command for closing a chest. This would only happen after the player had emptied all three and climbed into one of the chests. At which point, the afore-mentioned **Action**, which followed on from the standard **Action** of the command for climbing into that chest, would add that **Event** onto the list.

At the beginning of the final phase, the same animation would be played back as when the player's attempt to escape had failed. And a **Primary End Event** would be sent at the end of that animation. But a different **Secondary End Event** would be required, since the player would have succeeded this time. Thus, a seventh additional **Secondary Event** would be required. This **Event** would require a corresponding **Action**, which would respond to the captors leaving the cabin and the door open, for the player to complete the escape.

This penultimate **Action** would require an eighth additional **Secondary Event**. This would be the final **Event**, just as under the first system. And it would only be sent when the player had reached the open door and completed the escape. However, in order for this to happen, that **Event** would either be required to follow on from any existing **Secondary Proximity Event** assigned to that door. In which case, whatever **Action** responded to that **Proximity Event** would be required to send an additional **Primary Event** at the end of it. So that the final **Event** could follow on from that **Primary Event**. But that should happen anyway according to the principles of the second system.

Or this final **Secondary Event** could be that **Secondary Proximity Event**. Either way, you would require a final **Action** that would respond to this **Event**. And that **Action** would define what would happen once the player had completed the escape. So in total, you would require eight additional **Primary Events**, nine **Secondary Events** and ten **Actions**.

There is an example of how to implement this phase of the game using the second system for generating new **Events** in the **Event-Database Architecture** in Figures 4.18, 4.19 and 4.20. Notice how the **Actions** complement each other in pairs.

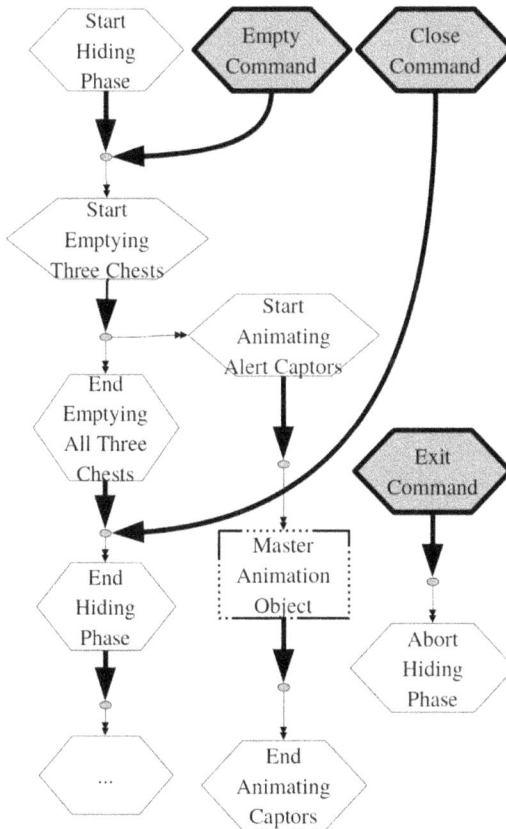

FIGURE 4.18 The same flow diagram as in Figures 4.15 and 4.16. Except there are seven new nodes or **Actions**. These allow for the captors to occasionally come in and disrupt the first or hiding phase of escape and manhandle the prisoner. As well as allowing the prisoner to exit the chest, after hiding in it, and abort the first or second phases.

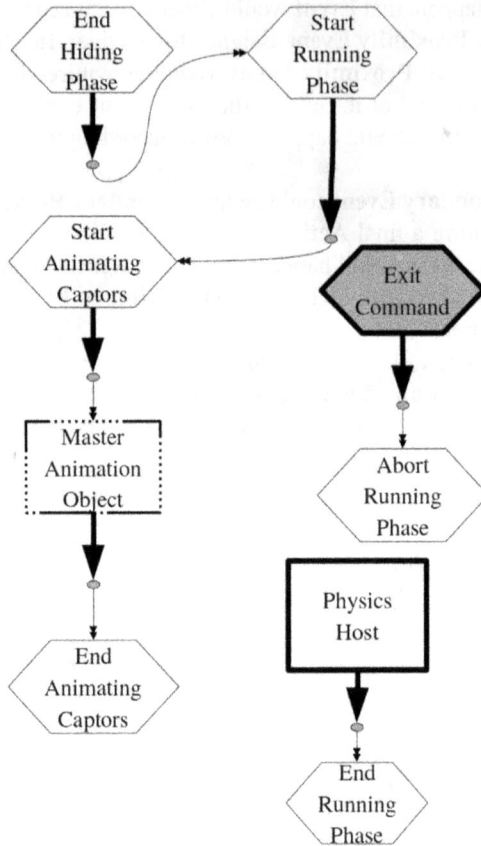

FIGURE 4.19   Extension of Figure 4.18.

The **Start Hiding Phase Action** is complemented by the **End Hiding Phase Action**. The **Start Emptying Three Chests Action** is complemented by the **End Emptying Three Chests Action**. The **Start Animating Alert Captors Action** is complemented by the **End Animating Captors Action**. The **Start Running Phase Action** is complemented by the **End Running Phase Action**. And the **Abort Hiding Phase Action** is complemented by the **Abort Running Phase Action**.

The contents of the **Game Database** would be more or less the same, under the second system, as under the first. That is, in so far as each Action would use the same *Records*, to display the same number of **Game Objects** in the *Game World*, count the number of chests that had been emptied and play back the same animations and sounds, as before. The difference would be that the number of **Primary Events** sent from within each **Action** would be greater. Although, in this particular instance, only eight of the additional **Primary Events** were used, to build this stage of the game, there would in fact be a lot more. In this example, there are 12 additional or Custom **Actions**, and each of them sends **Primary Events** at the beginning and end. So, in total, there are 24 **Primary Events** available. Of these, only eight are being

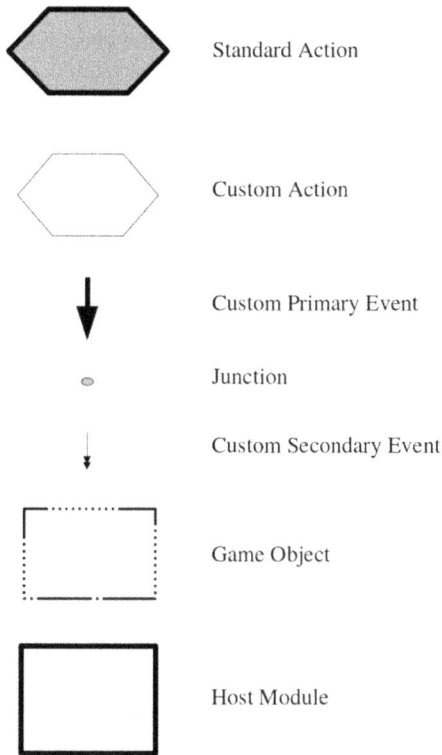

FIGURE 4.20 Legend of symbols displayed in Figures 4.18 and 4.19. It is a list of the symbols, for the components of the Event-Database Architecture, that could be used to build one stage of a game.

used, and those are the ones you see in Figures 4.18 and 4.19. But there are another 16 which are not shown and may never be used (Figure 4.20).

This difference between the first and second systems for generating **Events** would also be reflected in the contents of each **Action**. Under the second system, each **Action** would send a **Primary Event** when it began and ended, instead of when it found out that some significant transition in the *data* had occurred. These two **Events** would be sent regardless of the changes that had occurred in the **Game Database**, during that Action. A **Game Object** could no longer respond to, for example, an **Event** sent when a chest had been completely emptied, as supposed to being partially emptied. Nor could a **Game Object** respond to an **Event** sent when the player had entered and closed one of the chests, after having emptied all three. So each Action would simply have to modify the *Database* itself, whenever it noticed any significant changes had occurred.

For example, the **Action** which responded to the contents of a chest being partially emptied would be responsible for noticing when the chest had been completely emptied. And it would subsequently increase the total number that had been emptied. Once all three had been emptied, that same **Action** would change the flow of the game. Before, the player climbing into one of the chests would have had no effect. Neither would the player subsequently closing that chest had any effect. But, after all three chests had been emptied, the flow of the game would be changed by the same **Action**. So that the player climbing into the chest would, in turn, lead to a new **Action**. That new **Action** would in turn produce another change in the flow of the game. So that closing the chest would in turn lead to a new Action. That would play the animation of the captors entering the cabin.

Naturally, the opposite **Action**, which responded to the player climbing back out of the chest, would have to reverse this change. It would have to stop the automatic appearance of the captors once the lid was subsequently closed, if it detected that this would happen at the end of it. Instead, it would have to change the flow of the game back. So that closing any of the chests would not have any effect. This should not bring in the captors, since the player would not have completed the escape whilst standing back out in the cabin.

Under the first system for adding **Events** to the **Architecture**, a transition in the *data* held by the **Game Database** would cause a **Primary Event** to be sent. But under the second system, such a change would only cause a change in the chain of **Events**, which controlled the flow of the game.

In other words, under the first system, excluding the standard **Primary Events**, there would be no change in the *Records* describing the relationship between **Primary Events** and **Secondary Events**, from the beginning to the end of a game. But under the second system, there would always be a change in these *Records*, whenever there was any significant change in the *Database*.

Another major difference between the two would be that you could easily extend the beginning or ending of each **Action**, performed in response to an **Event**, under the second system. Suppose the *game design* were changed, so that a new sound would be heard before a particular chest was opened, or just after it had been opened. If that were to happen, the software could easily be edited to achieve this effect. This could be achieved by simply editing one of the **Game Objects**, to include a new Action, that would respond to the **Primary Event** sent before that chest was about to be opened. This would require a new **Secondary Event**, which would follow on from that **Primary Event**, which the Action would respond to. And, in response, it would play back this new sound. Of course, the new sound would have to be recorded and added to the *Records* of **Game Database**, along with the new **Event**, first.

Alternatively, suppose the *game design* were changed, so that the same chest could not be opened, unless the player had found some key earlier, lying around in another part of the *Game World*. This change too could easily be accommodated. One of the **Game Objects** could be edited, to include a new Action that would respond to the **Primary Event** sent after that chest had been opened. This would require a new **Secondary Event**, which would follow on from that **Primary Event**, and this Action would respond to. And, in response, it would simply check if the player's character were carrying the right key. It could do this by, for example, searching a *Record* holding the list of items in the inventory of that character. And if it could not find the right key, the new Action would immediately shut the lid. It could do this by sending the same **Primary** and **Secondary Events**, which would otherwise be sent, when the command for closing the chest had been used.

But there would be other, far less obvious, extensions you could make to the **Actions** performed in response to an **Event**, under the second system. Suppose you had a Action that would reset all the characters and other items involved in this stage, before it was played. And this Action responded to a **Primary Event** sent when that stage had begun. Each time it responded to this **Event**, it would use the same six **Actions**. The first of these would clear the list of **Game Objects** being displayed in the *Game World*. The second would add the cabin and its background scenery onto this list. The third would place the three chests inside the cabin. The fourth would place a locked door over the entrance of the cabin. And the sixth would reset the animations and sounds played back when the player moved, when the captors entered the cabin to start searching for the player, and when they reentered prior to manhandling the player.

Now each of these six **Actions** would be required to send two different **Primary Events**, when it began and ended, under the second system. So you could easily extend each one if say, the *game design* were changed so that the cabin would include an old clock, against one of its walls. And a table would be placed somewhere in the middle of the cabin. Furthermore, instead of the player's character beginning in one corner of the cabin, in one pose, that character would begin in the opposite corner in a different pose.

To make these changes, three new Actions could be added to the **Game Objects**, without editing any of the existing ones. And each one would require a new **Secondary Event**, following on from an existing **Primary Event**, that it could respond to. The first new Action could respond to the **Primary Event** sent after the cabin had been loaded in the *Game World*. And this could loaded the old clock inside it, against one of the walls. The second Action could respond to the **Primary Event** sent after the chests had been loaded inside the cabin. And this could search for an empty space into which it could place the table. Finally, the third Action could respond to the **Primary Event** sent after the player's character had been loaded in the cabin. And this could move that character, from one corner to the other, while changing the player's initial pose. (Note that the standard Architecture does not include a Primary Event for when items are Loaded into the Game World. But the Architecture described in the second book in the series, Event-Database Architecture for Computer Games: Game Design and Nature of the Beast, does include an example that does include this Primary Event).

Unlike the second system, the third system would be similar to the first system for adding **Events** to the **Architecture** in one respect. That is to say, it would also require a **Primary Event** to be sent only when some significant transition in the *data* had occurred. Under the third system, if a *software procedure* included some *logic branch* that checked for such a transition, it would have to send an **Event** when that transition had occurred. But this system would be much broader in this respect than the first. For that *procedure* would also be required to send another unique **Primary Event**, when no such transition had occurred. Furthermore, all *logic branches*, excluding those that were merely either checking for errors, or searching for *data*, would have to send two or more **Events**. And this would depend on whether some condition had been met or not.

So suppose, for example, you had to build this same stage of the game, where the player was being held captive in a cabin, and had to find a means of escape. Instead of the set of nine or ten **Actions** used to build that stage under the first, and the eight or nine for the second, under the third system you would require just six or seven.

You would require one **Action** that would respond to the **Events** sent when the player had partially emptied the contents of a chest. You would require another that would respond to the **Events** sent when the contents of a chest had been completely emptied. You would require another that would respond to the **Events** sent when the captors had to reenter the cabin and start manhandling the player. You would require another that would respond to **Events** sent when the initial phase of the escape had been completed. You would require another that would respond to the **Events** sent when the final phase began. And, finally, you would require one **Action** that would respond to the **Event** sent when the player had completed the escape.

In other words, two of the standard **Actions**, which would be required to perform the player's commands, would be required to include a *logic branch*. And these would in turn provide the additional **Primary Events**, which a solution could be built upon. These two **Actions** would namely be the one that would perform the emptying of each chest, and the one that would close each chest. The former would be required to send one additional **Primary Event**, when a chest had been completely emptied. And it would be required to send another, when it had only been partially emptied. The latter would be required to send one additional **Primary Event**, when the player had closed a chest whilst standing outside the chest. And it would be required to send another, when a chest had been closed after the player had climbed into one.

The first additional **Secondary Event** would be required to follow on from the **Primary Event** sent when a chest had been completely emptied. This **Secondary Event** would also follow on from the **Primary Event** sent when the player had climbed into a chest and closed the lid. This would require a corresponding **Action**, which would count how many chests had been emptied. The **Action** would use one *logic branch*, to send one of a trio of additional **Primary Events**, depending on whether all three chests had been emptied. Either it would send one, when all three chests had not been emptied. Or it would send another, when the **Action** had followed on from the command for closing a chest. Or, it would send another, when the **Action** had followed on from the command for emptying the chest.

To determine which command the **Action** was following on from, the **Game Objects** were performing that **Action**, could look up the **Events History Record** to determine preceding **Primary Events**. This would include the standard **Primary Events** for each command.

Nevertheless, this would not matter for the last of this trio of **Primary Events**, sent by the **Action**, in this example. Since this would have no subsequent **Secondary Events**, nothing else would happen once the player had emptied all three chests. The first one, however, would be followed by an additional **Secondary Event**. This **Event** would also follow on from the **Primary Event** sent after a chest had only been partially emptied by the player's command. This **Secondary Event** would require a corresponding **Action**, which would occasionally alert the captors to the player's escape. And it would play back the animation of the captors rushing in to begin a frantic search.

This **Action** would behave in the same way as the one under the first and second systems. It would require a third additional **Secondary Event** that would act as the **Secondary End Event** of the animation. This **End Event**, in turn, would require another **Action**, which would play back the animation of the captors reentering to manhandle the player.

As for the second of the trio of **Primary Events**, sent by the **Action** which would detect when all three chests had been emptied, this would mark the completion of the initial phase of the escape. Since this **Event** would follow on from a command to close the chest, after the player had entered it. So this would require a fourth additional **Secondary Event** and a corresponding **Action**, which would respond to the end of this phase. This **Action** would play back the animation of the captors briefly entering the cabin and then leaving to begin a frantic search. As such, it would behave in very much the same manner, as the one under the first and second systems.

Once the animation had been played, the **Action** would require a fifth additional **Secondary Event**, which would act as a **Secondary End Event** of the animation. This would require a corresponding penultimate **Action** that would determine when the player had completed the escape from the cabin. This would be, that is, when the player had reached the door. So either whatever **Secondary Proximity Event** assigned to the door would be required to send an additional **Primary Event**. And the **Action** would append its **Secondary Event** onto that **Event**. Or its **Secondary Event** would be used as the **Secondary Proximity Event**. Either way, this **Event** would require a corresponding final **Action**. And that **Action** would define whatever rewards the player would receive, or subsequent stages of the game that would be followed. Therefore, in total, six or seven **Primary Events**, with six or seven subsequent **Secondary Events** and corresponding **Actions** would be required (see Figures 4.21–4.23).

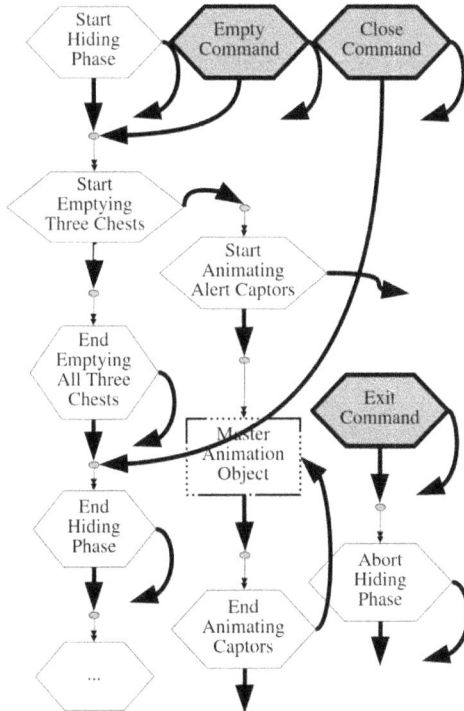

FIGURE 4.21 An example of Primary Events, Secondary Events and Actions that could be generated using a simple rule or system. That in turn could be used to build one quest of a game. This quest being trying to escape from captivity in a cabin, with three chests full of treasure, on board a pirate's ship.

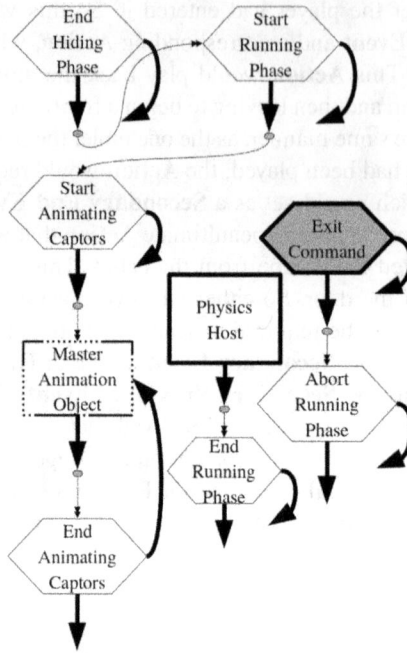

FIGURE 4.22 Extension of Figure 4.21.

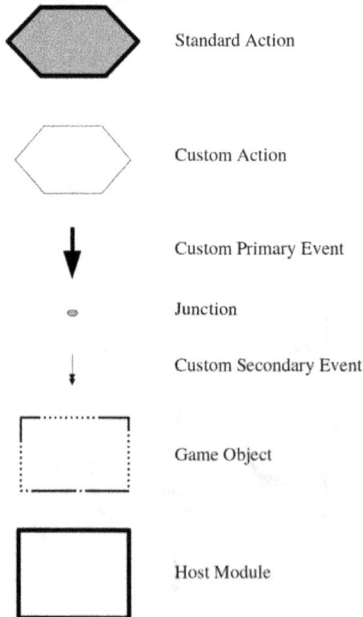

FIGURE 4.23 Legend of symbols displayed in Figure 4.21. It is a list of the symbols, for the components of the Event-Database Architecture, that could be used to build one stage of a game.

Under the third system for adding **Events** to the **Architecture**, every **Primary Event** added would have at least one other complementary **Event**. For example, the **Events** sent when the player had partially emptied the contents of a chest would have complements, in the **Events** sent when the player had completely emptied that chest. Similarly, the **Events** sent when the player had closed the lid, from inside a chest, would have complements in the **Events** sent when the lid had been closed from the outside. Likewise, the **Events** sent when the player had successfully completed the initial phase of the escape would have complements too. These would include the **Events** sent when that phase had failed and the captors had noticed the cargo being offloaded. And these would also include **Primary Event** sent when the player had just emptied all three chests.

During a production process which followed the **Event-Database Architecture**, each set of complementary **Events** would be added together, to the **Game Database**, at the same time. This would be because each set would be used by the same **Action**. And, according to steps (7) and (17) of the **Event-Database Production Process**, each should be added before that **Action** was written. So anyone who came across each set, such as the *Game Designers*, would also see the options that the complementary **Events** gave them, to modify the *game design*.

For example, the last set of complementary **Events**, relating to the success or failure of the initial phase of the escape, would present the *Game Designers* with at least one obvious possibility. This being namely that the captors could enter the cabin at any time, during the player's escape, and foil the plan. They need not wait to enter, once the player had climbed into one of the chests and closed the lid. The **Action** that responded to the **Events** sent after the player had closed the lid, with one of a trio of **Primary Events**, could be used at any time. And depending on whether the player had successfully completed this phase, or not, it would send one of its complementary **Events**.

So they need not predicate the entrance of the captors, on the commands for emptying and closing a chest. Instead, they could simply have the captors appear after a random interval, from the start of the stage, had passed. This change could be easily achieved by stopping all the **Secondary Events**, which would follow on from the **Primary Events** sent by the standard **Actions** of these commands, from doing so. Instead, the **Secondary Event** that would have been sent after the player had closed the chest could be added to the list of delayed **Events** sent when the stage had begun. And the properties of this **Secondary Event** could be edited to include a time delay, proportionate to the minimal interval before the captors' entrance. All of this could be done by editing the **Game Database**.

Furthermore, the **Action** that would have responded to the **Events** sent when the player had failed, to complete the initial phase of the escape, could be edited. So that it would no longer occasionally alert the captors to the escape. Instead, if by the time the delayed **Event** had occurred, the player had not completed the initial phase of the escape, it would always alert the captors.

Now the escape from the cabin would become a tense affair. The captors could return at any moment to foil the player's plan. It could take several attempts, at this stage, before the player would succeed. Alternatively, it may take a long time before the captors enter the cabin, after the player had successfully completed the initial phase of the escape. As a result, the player may become impatient and step out of hiding too soon, only to be caught by the captors.

The tension could be heightened even further by adding a new sound, and **Secondary Event** to the **Game Database**, along with a new **Action** to the **Game Objects**. The new sound would be a recording of footsteps outside the door of the cabin that would be heard prior to each entrance of the captors. The new **Secondary Event** would be sent, after a suitably long interval from the start of the stage, but just before the captors appeared. And the new **Action** would respond to this **Event** by playing back the recording. So that the player would be given advance warning every time the captors were about to enter.

As has been illustrated in this example, the tolerance for changes to a *game design* would compensate for the greater number of *software procedures* (i.e. **Actions**) that would be required. Be it under the first, second or third systems for adding **Events**, to the **Architecture**, that number could be up to ten times greater than under the *Software Evolution Process*. But it would be a mistake to view the **Architecture** only from this lower level, the level of the *technical design*; thereby exaggerating this effect. A higher view, from the level of the *game design*, which the **Architecture** was meant to serve, would provide a better or more accurate perspective.

It would be easy to exaggerate, from a low-level view, the number of *software procedures* required during the production of a game, by making false assumptions. Namely, that the relationship between the number of *procedures*, required under the **Event-Database Architecture**, to the number required under the *Software Evolution Process*, would be linear. This would be the case at the beginning of production. That is to say, the number of *software procedures* or **Actions** generated at the beginning of the **Event-Database Production Process** would be the same if not greater than the number of *software procedures* generated at the beginning of a *Software Evolution Process*. The rules for generating **Events** in the **Event-Database Production Process** require you to generate lots of redundant **Events**, and hence *software procedures* or **Actions**, to respond to those **Events**.

But as time passed, and the number of **Events**, **Actions** and **Game Objects** grew in the **Architecture**, the combinations of **Events**, **Actions** and **Game Objects** you have available to build the game, just by editing the **Game Database**, would grow exponentially. And only a subset of these combinations would be required to replicate the same features of a *game design* that would otherwise be performed by all of the *procedures* built under the *Software Evolution Process*. And towards the end of production, the combinations available to meet hypothetical changes in the *game design* would exceed all expectations. All of these changes could be accomplished by the *Game Designers* or other staff, without the aid of the *Game Programmers*. And the majority of these same changes could not be replicated without a major overhaul of the *software procedures*, under a *Software Evolution Process*. For most of these *procedures* would not have been documented. Those that were would only be understood, at worst, by the *Programmers* who wrote each one, and at best, by other *Programmers*.

Likewise, it would be easy to exaggerate the number of **Events** you would require, for example, under the first system for adding **Events**. This system would not be as broad as you might think. During a production process, you would not add **Events** based on every transition of *data* that occurred on the level of the *technical design*. Instead, you would add one for every transition that occurred on the level of the *game design*. An example of when you would send an **Event** using this system would be when a player used a significant command. You would also send an **Event** when the player came within close proximity of a special **Game Object**, in the *Game World*. When the number of items being carried by a character, crossed an upper or lower limit, would cause another **Event** under this system. When some modified property of a location, or an item lying in that location, matched another, constant *Field* in the **Game Database** would also produce an **Event**.

Nevertheless, the second and third systems, for adding **Events**, would be both flawed. The second would ignore internal *logic branches*, which may occur within a *software procedure*. The third would ignore the external *branches*, from one *procedure* to another. Both of these *branches* may be of interest when it came to changing the *game design*. A solution to both of these problems would be to combine the second and third systems, to form a fourth system i.e.

4. two unique **Primary Events** should be sent at the start and the end of each *procedure*, or **Action** and each option, chosen by a *logic branch* within that *procedure*, should be performed by another *procedure* or **Action**.

This would provide you with the opportunity to not only easily extend both ends of a *software procedure*, or **Action**, just by editing the *Database*. But you may likewise extend any options, chosen by the *logic branches* that occurred internally, within it, as the changing of the *game design* dictated.

4.3.5 APPLICATION: GAME PLAY – PICKING A ROSE BUSH

Compare how the two systems for generating **Game Objects** mentioned in the subchapter entitled **Rules for Generating the System of Game Objects** would be used in a production process, based on the **Event-Database Architecture**, with the *Software Evolution Process*.

For example, suppose one stage of a game was set in a courtyard of some medieval castle. And in this courtyard was a garden, with one giant rose bush at the centre of it. If the player stayed long enough in that garden, he or she could watch the roses growing on the bush. Each rose would first appear as a small bud, which would then grow larger, open slightly, blossom and finally die. But the player could pick a rose once it had blossomed. And once picked the rose would permanently remain blossomed. However, if the rose died on the bush, the whole cycle would begin again, with the appearance of a small bud.

In this example, a rose would have five states. But the rest of the bush would only have one state. So you would require one **Game Object** to build the roots, branches and leaves of the bush. And you would require five **Game Objects** to build each rose, under the first system for adding **Objects** to the **Event-Database Architecture**. One **Game Object** would be required for each state. When the first one appeared in the *Game World*, it would send one unique **Primary Event**, followed by two subsequent **Secondary Events**, delayed by a few minutes. In response to one of these delayed **Events**, that **Game Object** would then disappear. And in response to the other delayed **Event**, the next **Game Object** would appear in its place. When the second one appeared, it too would wait a similar amount of time, before disappearing just like the first, and being replaced by the third **Game Object**, and so on. Finally, after the last one had appeared, it would bring the first one back up in the same fashion, and the whole cycle would begin again.

The fourth **Game Object** of this cycle would represent a rose that had blossomed. And it would require one unique **Primary** and **Secondary Event** to respond to, when it was picked by a player. When this happened, the flower would disappear from the bush, appear on the body of the player's character and stop waiting to die. That is to say, the **Game Object** would remove the **Secondary Event** it was due to receive, when it was time for the rose to die, from the *Record* holding the list of delayed **Events**.

Clearly, if the bush had several roses, and the state of each rose was independent of the others, then you would require more **Game Objects**. You would require five **Game Objects** for each rose. And this would apply for any feature of a game you wanted to implement. If the original single **Game Object** you wrote to implement the feature had many elements, which had independent states, you would require more **Game Objects**. You would first have to write a single **Game Object** for each element. And, after that, you would replace each of these with multiple new **Game Objects**: one for each state.

With this system, every **Game Object** would represent one element of the game, in one state. This element could be a character, a location or some other item found in one of its stages. And you could predict how each **Game Object** would behave in this state, in response to any **Events**. So anyone could combine these **Game Objects** with other elements of a game, to produce the effect he or she wanted.

Contrast this with how the same example would be built in a *Software Evolution Process*. The rose bush would be evolved over several steps. The first step would involve displaying a static image of the roots, branches and leaves of the bush, in the *Game World*. This may also include some static roses at different stages of growth. The second step would involve demonstrating a single rose going through a single cycle of growth; from a small bud, to a fully blossomed rose, and finally ending with its death. The third step would involve showing the rose repeating this cycle, again and again, ad infinitum. The fourth step would involve several roses repeating this same cycle, in unison, on the bush. The fifth step would have these same roses, repeating this same cycle, but this time out of step with each other, and at different phases. The final step would introduce the player's character. And this would include the commands for moving that character around, as well as picking a rose once it had blossomed.

It all sounds reasonable, even desirable. Until you bear in mind that, as would be typical of a *Software Evolution Process*, each step would be executed in an ad hoc manner. And there would be no regard for any subsequent steps. One result of this would be that there would be no system for deciding how many *game modules* would be used to build the rose bush.

In the worst case, you may end up with only one *module* being used. As the *game design* changed, that single *module* would end up becoming very complex. No attempt would be made to simplify it, by breaking it down into smaller *modules*. Instead, this single *module* would be allowed to grow larger and larger, as more and more features were added to the rose bush.

This *game module* would display all of the roses and the rest of the bush. It would contain all of the *data*, including those describing the position of the roses and the rest of the bush. It would include the *data* describing the different *Frames* of animation, used to display each rose, during its growth. It would include the *data* describing the stage of growth of each rose. It would include *data* describing how much time had passed since the rose bush appeared in the *Game World*. And the *module* would include the *software procedures* used to read and react to the command, from the player to pick up a rose, once it had blossomed. And, of course, none of these *data* and *procedures* would be documented.

Along with its over complexity and its lack of documentation, the *game module* used to build the rose bush would not be re-usable. In the worst case, it would have been tailored only for this one particular *game design*. It could not be used to subsequently display other plants, with other flowers going through different stages of growth. Nor could it even be used to display the same rose bush, with more roses.

In the best case, the rose bush would be built using two *game modules*. One would be used to build a single rose, which would also be re-used to build all the roses. And another would be used to build the roots, the branches and the leaves. And you could edit the images or 3D models used to display the roots, branches, leaves and roses to change their appearance, in a *game-editor* and re-use these to display other plants going through different stages of growth. Although this would not suffer from the same over complexity as the previous case, the difference would be marginal. The number of flowers would be hardcoded. The number of states of these flowers would be hardcoded. The command to pick a flower would be hardcoded. You could not edit these items by simply editing a **Game Database**. You cannot edit these while the game was being played. You cannot edit these and restart the game. You cannot edit these without *Programmers* or programming tools rebuilding the **Game Database**. You may be able to edit these and preview the effect in the *game-editor*. But the *game-editor* is not the game. A preview is not a final product. A preview of a film is just marketing. Whereas before, in the worst case, you had one *module* rapidly growing out of control, now you would have two *modules*, slowly growing out of control. Although there may have been an initial attempt to simplify the software, it would have only been discretionary. Subsequent attempts would forego such niceties, when the *game design* was changed. And any additional features would be crammed into the same two *modules*.

Furthermore, although there may be some documentation of the original two *game modules*, this too would also have been discretionary. There would be no further documentation of the *modules*, after the initial *game design* had changed, and the *software procedures* and *data* changed along with it. So instead of having no information, as in the worst case, you would have misleading information, in the best case, when it came to editing the software.

Finally, although these two *game modules* may be re-used to, say, build other plants in the *Game World*, you would eventually find this capability also to be misleading. For either, in order to do this, you would be required to edit a third *module* which had been neglected and as a result had grown out of control, become very complex, had no documentation and its relationships with the two *modules* was also not documented. Or you would find out that this third *module* was in fact part of the *game-engine*. And so, by changing it, you would be putting other *modules*, which depended on this *game-engine*, at risk. Or you may find access to the *game-engine* denied, for this very reason, by the *Programmers* who wrote it or maintain it. And so you would have to resort to finding some crude, ad hoc solution to get around this limitation.

However, with the **Event-Database Architecture**, there would be no such limitations. At least with respect to the *data*, these would be documented in the *data design*. And all of the *software modules*, including the software components of the **Architecture**, such as the **Events Host** and **Database Host**, could be edited. So long as these components kept to the minimum requirements described for each one and followed the principles of the **Architecture**. Any of these could be edited if that were required by the *game design*. This would be some of its advantages over a *game-engine*.

Another would be that the *game-engine*, used in a *Software Evolution Process*, would probably have been produced by another *Software Evolution Process* anyway. Along with the other tools, used to give the process credibility at its onset, such as the *game-editors*, the *game-engine* would be suffering from the same flaw. All of these tools would have been produced by another *Software Evolution Process*. And as such these tools would not be as well documented, comprehensible and accessible as a *software architecture*. Any discussion involving these tools would invariably degenerate into an esoteric discussion about the idiosyncrasies of these tools, because of the same degenerative language, because of this same flaw inherent within each product of a Software Evolution Process. And thus the discussion has to be either left to the *Programmers* who wrote the tools. Or it has to be left to those who could understand this degenerative language, that is to say other *Programmers*.

But, by following the **Event-Database Architecture**, the level of discussion, during a software production process, about any aspect of a game, would always remain high. It would not need to go any lower than **Events**, **Actions**, **Game Objects**, *Database Tables*, *Fields* and *Records*. So hopefully everyone would always be able to participate: not just the *Game Programmers*. This accessibility would be evident under any system you chose for adding **Game Objects** to the **Architecture**.

For example, under the second system, you would begin building a feature for a game by deciding what formulas you were going to use. These would not be formulas which revolved around mathematics, physics or some other similar science. Nor would these be formulas which revolved around programming languages, which only the *Game Programmers* could understand. Instead, these would be formulas which simply revolved around the *Tables, Fields* and *Records* of a **Game Database**.

There would be two types of formulas. The first type of formula would be a linear sequence of calculations, or operations. Either the formula may use some information (i.e. *Fields* in the **Game Database**) to produce a result (i.e. changes to other *Fields* in the *Database*). Or the formula may use no information at all and simply repeat an operation (i.e. modify one or more *Fields* in a prescribed way) over and over again. In either case, each step of the formula would always be performed. The second type of formula would simply check a set of *Fields* in the **Game Database**. And it would send a **Primary Event** when these *Fields* matched certain criteria.

Once you had decided the formulas you were going to use, to build the feature of a game, you would write one **Game Object** to evaluate each one. Each **Game Object** would respond to one or more unique **Secondary Events**. And when it received an **Event**, the **Game Object** would use the same formula.

For example, suppose you had another stage in a game, where the player had to enter a sealed room, in a medieval castle. But when the player had reached the entrance of this room, he or she would find it guarded by a wizard. And, in order to get past the wizard, the player would be told to find at least five, out of a possible ten, magical items, hidden around the rest of the castle. Once the player's character had collected these items, that character could enter the room and move on to the next stage.

Now under the second system for producing **Game Objects**, you would require one *Field* in the **Game Database** and 22 formulas, to build this stage. The *Field* would be required to hold the total number of items which the player had picked up. One formula would be required to reset this total to 0, when the stage began. Ten formulas would be required to increase the total by one, when the player found an item. Ten formulas would be required to decrease the total when the player lost an item. And one formula would be required to send a **Primary Event** when the total was greater than five, to indicate that the stage was over.

So to implement this stage of the game, 22 **Game Objects** would be required: 1 for each formula. And these **Game Objects** would need **Events** to indicate when the stage had begun, when the player had picked up each item, when the player had dropped each item and when it was time to check how many items the player had collected.

It may be tempting to use less **Game Objects**. Since the ten formulas used when the player picked up an item would be the same, you may be tempted to use one **Game Object** to implement those ten formulas. Likewise, you may be tempted to do the same for when the player dropped an item. If this stage of the game were built under the *Software Evolution Process*, it would certainly succumb to such temptation. That is to say, one *game module* would be used to read and respond to the command to pick up each item. That same *module* would be used to count how many items the player had picked up. That same *module* would determine when the stage was over. And if all of these magical items had the same appearance, then that same *module* would be used to display all of these items.

But, with the **Event-Database Architecture**, you would need at least ten **Game Objects**, because the items would have different sets of properties, such as the position of each one. This would anticipate any changes to the appearance or the other properties of each item, which could conceivably happen. Furthermore if, for some reason, the *game design* were changed, and different weights were given to picking up one item, than another, then you would need two different formulas for each item. No two formulas used when an item was picked up would be the same. Likewise, no two formulas used when an item was dropped would be the same. So you would require all 20 **Game Objects** for the ten items.

With this system, every **Game Object** would represent one formula used by one element of the game. And, like the first system for producing **Game Objects**, you could predict how each **Game Object** would behave in response to any **Events**. So anyone could combine the different formulas of the game to produce a desired effect.

4.4   STEP 4: DATA DESIGN

The *data design* would be basically a description of the **Game Database** or a *Relational Database*. That would contain all of the *data* that would be used by the game, including public or shared *Game Data* as well private or restricted *Abstract Data*.

This would include a description of all of the *Database Tables*, columns or *Database Fields*, rows or *Database Records* in each *Table*.

This would be designed by the *Database Administrator* after consulting the *Game Programmers*, *Game Artists*, *Sound Designers*, *Game Designers* and *Game Testers*.

The *Game Programmers* would be consulted to find out what *Game data* and *Abstract Data* they needed to implement the initial features of the *game design*. Using whatever system of **Events** and system of **Game Objects** had been chosen for the **Event-Database Architecture**. This includes implementing common or popular features such as simulating physics, displaying 2D or 3D graphics and animations, or implementing Artificial Intelligence. This includes any *data* required by the **Host Modules** of the **Architecture**.

The *Game Artists* would be consulted to find out what *Game data* they required to build 2D or 3D graphics and animations.

The *Sound Designers* would be consulted to find out what *Game data* they required to store and play back sound effects or music or mix sounds at different points of the game. The *Sound Designers* would also be consulted about any data related to **Events** added to the game. To make sure that the names of the *Database Records* for these **Events** used to implement a feature were consistent. And they could query the *Database* to find all the **Events** related to same feature, which they had to add sound for, using the same name.

Likewise, the *Game Designers* and *Game Testers* would also be consulted about any *data* related to **Events** added to the game. To make sure that the names of the *Database Records* for these **Events** used to implement a feature were consistent. And they could query the *Database* to find all the **Events** related to same feature, which they had to either modify or test, using the same name.

4.5 STEP 5: TOOLS DESIGN

A *tools design* would be a basic description of the custom and third-party tools that would be required by the *Database Administrators*, *Game Programmers*, *Game Artists*, *Sound Designers*, *Game Designers* and *Game Testers* to build a game with the **Event-Database Architecture**. You can see an example of it in Table 4.2.

TABLE 4.2

Table of Third Party and Custom Tools for the Archetypal Game Based on the Event-Database Architecture

Name	Marketing Description	Staff	Copies
LibreOffice version 6.0.7.3	LibreOffice is a modern, easy-to-use open-source productivity suite for word processing, Spreadsheets, presentations and more.	All	65
GCC 7.5.0	The GNU Compiler Collection includes front ends for C, C++, Objective-C, Fortran, Ada, Go and D, as well as libraries for these languages (libstdc++, …). GCC was originally written as the compiler for the GNU Operating System. The GNU system was developed to be 100% free software, free in the sense that it respects the user's freedom.	Game Programmers	20
GIT 2.17.1	Git is a fast, scalable, distributed revision control system with an unusually rich command set that provides both high-level operations and full access to internals.	All	65
glibc 2.27-3	The GNU C Library – The project provides the core libraries for the GNU system and GNU/Linux systems, as well as many other systems that use Linux as the kernel. These libraries provide critical APIs, including ISO C11, POSIX.1-2008, BSD, OS-specific APIs and more. These APIs include such foundational facilities as open, read, write, malloc, printf, getaddrinfo, dlopen, pthread_create, crypt, login, exit and more.	Game Programmers	20
libc++ 3.4.25	libc++ is a new implementation of the C++ standard library, targeting C++11 and above.	Game Programmers	20

(Continued)

TABLE 4.2 (*Continued*)
Table of Third Party and Custom Tools for the Archetypal Game Based on the Event-Database Architecture

Name	Marketing Description	Staff	Copies
MESA 20.0.8	Free implementation of the OpenGL API – GLX vendor library.	Game Programmers	20
libsdl2 2.0-0	Simple DirectMedia Layer is a cross-platform development library designed to provide low-level access to audio, keyboard, mouse, joystick and graphics hardware via OpenGL and Direct3D.	Game Programmers Sound Designers	22
XBox SDK 1.00	Software library for developing games on the Microsoft XBox Game Console.	Game Programmers Sound Designers	22
PS5 SDK 1.00	Software library for developing games on the Sony Playstation 5 Game Console.	Game Programmers Sound Designers	22
libmysqlclient 5.7.42	Software library for developing MySQL Clients.	Game Programmers	20
libopenal 1:1.18.2-2	OpenAL is a cross-platform three-dimensional audio API. The API's primary purpose is to allow an application to position audio sources in a three-dimensional space around a listener, producing reasonable spatialisation of the sources for the audio system (headphones, 2.1 speaker output, 5.1 speaker output etc.).	Sound Designers	2
GDAM	GDAM is a digital DJ mixing software package. It aims to be a powerful, professional-quality music mixing and remixing system, suitable for live performance.	Sound Designers	2
CDex	CDex can extract the data directly (digital) from an Audio CD, which is generally called a CD Ripper or a CDDA utility. The resulting audio file can be a plain WAV file (useful for making compilation audio CDs) or the ripped audio data can be compressed using an audio encoder such as MP3, FLAC, AAC, WMA or OGG.	Sound Designers	2
Blender 2.79	Blender is the free and open-source 3D creation suite. It supports the entirety of the 3D pipeline—modelling, rigging, animation, simulation, rendering, compositing and motion tracking, even video editing and game creation.	Game Artists	40

(Continued)

TABLE 4.2 (*Continued*)
Table of Third Party and Custom Tools for the Archetypal Game Based on the Event-Database Architecture

Name	Marketing Description	Staff	Copies
Gimp 2.8.22	GIMP is a cross-platform image editor available for GNU/Linux, macOS, Windows and more Operating Systems. It is free software, you can change its source code and distribute your changes. Whether you are a graphic designer, photographer, illustrator or scientist, GIMP provides you with sophisticated tools to get your job done.	Game Artists	40
MySQL 5.7.42	The MySQL software delivers a very fast, multi-threaded, multi-user and robust SQL (Structured Query Language) database server. MySQL Server is intended for mission-critical, heavy-load production systems as well as for embedding into mass-deployed software.	Database Administrators	1
Database Host Query Custom Tool	Queries the **Game Database** for Tables, Records and Fields matching criteria.	All	65
External Events Host Custom Tool	Manually fires Primary and Secondary Events.	Game Testers	2

An example of the third-party tools has already been mentioned in the previous chapters

Documentation Tools
Programming Tools
Software Libraries
Art Tools
Database Tools
Sound Tools.

The *tools design* would also include a description of the custom tools that the staff would require.

Database Host Query Custom Tool would be a tool which allows the staff to query the **Game Database**, to find *Database Tables*, *Records* or *Fields* which match a certain criteria and display their contents. This includes *Database Records* for **Events** or **Game Objects** with a particular name which match the name of a feature of the game implemented using those **Events** or **Game Objects**. So that the *Game Programmers*, *Game Designers* and *Game Testers* could find out information about features. Or modify or test features by modifying the *Database Records* of **Events** or **Game Objects** used to implement those features.

The *Game Programmers* could use this to find any *Bugs* related to some feature. By examining all of the **Events** or **Game Objects** used to implement that feature, and their *Database Records*.

The *Game Designers* could use this to extend a feature by finding the chain of **Events** used to implement that feature, and changing the chain of **Events**. Or they could change the parameters of a feature by changing the *Database Fields* of the *Database Records* of the **Game Objects** used to implement that feature.

And the *Game Testers* could use this to exhaustively test a feature. By finding all of the *Database Records* of the **Events** and **Game Objects** used to implement it. And manually firing the **Events** through another custom tool. Or changing the parameters or *Database Fields* of the **Game Objects**.

External Events Host Custom Tool would be a tool which *Game Testers* and other staff could use to manually fire **Primary Events** and test the **Secondary Events** and **Actions** that are executed in response to it. There is a complementary tool called **Internal Events Host Custom Tool** which allows the players to do the same. This is described in the chapter 1.5 **Step 5: LPmud Tools Design** in the book, *Event-Database Architecture for Computer Games: Game Design and the Nature of the Beast.*.

NOTES

1. *Data design.* A description of all the data needed by a game. It is also a description of all the data produced by the tools used to build a game.
2. *Tools design.* A description of all the tools used to build a game. These include the tools used to create the data, to write the computer files used to build the software, to process the data or to archive the data and the computer files.
3. *The Software Architecture.* The description of all the software modules that would be required to completely implement an **Event-Database Architecture**. See the chapter entitled The Software Architecture.
4. *Entity-Relationship diagram.* A diagram which shows all the items (or entities) stored in a Relational Database, and the relationship between these items.
5. *Basic Set theory.* A branch of mathematics concerned with producing rational conclusions, about items in the real world, by only operating with these in groups. These operations can be used to build any computer hardware or software, including Relational Databases and software procedures.
6. *Hash-Table.* A table of information where the entries have been positioned using a Hashing Function. A Hashing Function is a software procedure which tries to map any random set of numbers or words onto a non-overlapping set of numbers, within a limited range.
7. *Professional Database sources.* Database Systems: Concepts, Languages, Architectures by Paulo Atzeni, Stefano Ceri, Stefano Parabosci and Riccardo Torlone.
8. *Data structure sources.* Introduction to Algorithms, second edition, by Thomas H. Cormen, Charles E. Leiserson, Ronald L. Rivest and Clifford Stein.
9. *Applied mathematics sources.* Vectors in Two and Three Dimensions (Modular Mathematics S.) by Ann Hirst.
10. *Physics sources.* Computational Dynamics, second edition, by Ahmed A. Shabana.
11. *Mathematics sources.* The Geometry Toolbox for Graphics and Modeling by Gerald Farin and Dianne Hansford.

12. *Computer graphics sources.* Computer Graphics: Mathematical First Steps (c) 1998, Prentice Hall. Patricia Egerton and William Hall.
13. *Sound engineering sources.* The DSP Handbook by Andrew Bateman and Iain Paterson-Stephens.
14. *Free software.* Software which can be freely copied, redistributed or modified according to its GNU Public Licence. The software comes with the computer files used to build it. So that the software can be easily modified.
15. *Electronic documentation sources.* OpenOffice.org 1.0 Resource Kit by Solveig Haugland and Floyd Jones.
16. *Programming sources.* The C Programming Language, second edition, by Brian W. Kernighan and Dennis M. Ritchie. The C++ Programming Language, by B. Stroustrup.
17. *Compiler sources.* Using GCC: The GNU Compiler Collection Reference Manual for GCC 3.3.1 by Richard M. Stallman and the GCC Developer community.
18. *Revision Control Software.* A tool used to store, retrieve, log, identify and merge different versions of software in production. It stores all the sources which produced each version e.g. The documentation of the software designs, the computer files used to build the software modules, the software data etc.
19. *Revision Control sources.* Essential CVS by Jennifer Vesperman.
20. *File library sources.* The Standard C library by P. J. Plauger.
21. *RDBMS resources.* Using SQLite: Small. Fast. Reliable. Choose Any Three. Paperback – Illustrated, 3 Sept. 2010.
22. *Graphics library sources.* OpenGL Reference Manual: The Official Reference Document to OpenGL, version 1.2 by the OpenGL Architecture Review Board.
23. *Audio library sources.* OpenAL Programming Guide © 2006, Charles River Media. Eric Lenyel.
24. *Multimedia library sources.* Focus on SDL by Ernest Pazera. Programming Linux Games – Building Multimedia Applications with SDL, OpenAL and Other APIs© 2001. No Starch Press. Loki Games. John R Hall.
25. *Computer Aided Design sources.* The Art of 3-D: Computer Animation and Imaging by Isaac V. Kerlow.
26. *Digital imaging sources.* Grokking the GIMP by Carey Banks.
27. *Relational Database Management sources.* MySQL: The Complete Reference by Vikram Vaswani.
28. *ASCII.* American Standard Code for Information Interchange. A common character set used in the US and UK computers.
29. *SVG Format.* Scalable Vector Graphics Format. A data format used to store images in a file, and to display images on the World Wide Web.
30. *FBX Format.* Film Box Format. A data format used to store 3D models, animations and associated digital data in a file, and display 3D models and animations in applications, developed by Autodesk.
31. *XML Format.* Extensible Markup Language Format. A language for describing other languages that describe structured documents stored in a file (e.g. a thesis, an article, a User Manual). It was designed to be flexible enough to store and display the huge array of documents on the World Wide Web. But its flexibility means there can be big differences in how it is used between any two documents.
32. *JSON Format.* JavaScript Object Notation Format. A data format for describing hierarchical data structures in a programming language called JavaScript. It was designed to store documents in a file, and to display documents on the World Wide Web.
33. *CSV Format.* Comma-separated Format. A data format for describing a Relational Database Table where each row in the table is represented by a line in a file. And the columns in the table are represented by words on each line separated by commas. So for example a 4 x 3 Database Table would have each row in the table represented by three lines. And on each line the entries in each column would be represented by four words separated by three commas.

34. *Newline character.* An ASCII character which marks the end of a line of text, and the beginning of the next.
35. *Escape character.* An ASCII character which is reserved for transforming the normal interpretation of a following character in a word. It is normally used to transform a sequence of characters into commands which control how text is displayed.
36. *X PixMap Format.* A data format used to hold images displayed on Graphical User Interfaces of computers that use the X Window System.
37. *CSV Format sources.* UNIX (TM) Relational Database Management by Rod Manis, Evan Schaffer and Robert Jorgensen.
38. *SVG Format sources.* SVG Essentials© 2002, O'Reilly Media. J. David Eisenberg.
39. *X PixMap Format sources.* X Pixmap© 2010, Beta Publishing. Lambert M. Surhone.
40. *Digital Audio Tape.* A magnetic tape used to digitally record music or computer data.
41. *Secure Digital Card.* A small portable flash memory card or microchip that stores data in a computer memory, up to 2 Gigabytes in size.
42. *Secure Digital High Density Card*. A Secure Digital Card that can store up to 64 Gigabytes of data in computer memory.
43. *Pulse Code Modulation.* A method of encoding an analogue signal in a digital data format. The signal is sampled at a constant rate, and the amplitude at each interval is converted into a number within a limited range.
44. *Digital recording sources.* Desktop Audio Technology by Francis Rumsey.
45. *Digital playback sources.* Modern Recording Techniques by David Miles Huber and Robert Runstein.
46. *Checksum.* A value that represents the total value of a series or sequence of data. That is used to check when there is an error in that series or sequence, when it is transferred from one computer or storage media to another.
47. *Logic path.* Any one of a finite, distinct sequence of actions (or instructions) that can be performed with (or within) a software system (or its software procedures).

5 Limitations or Criteria for Use

Many software projects go over schedule because the projects attempt to extend the software beyond its original limitations. At which point, the software becomes part of the problem, instead of the solution. Although the **Event-Database Architecture** attempts to mitigate this very problem, the **Architecture** could just as easily become part of it. Thus, it would be important to understand the limitations of the **Architecture**.

These limitations are defined by the problem the **Architecture** attempts to solve. This has already been described, but here is a summary:

1. The **Architecture** would not be beneficial to projects where the *game design* was complete. It would not be beneficial to projects which followed the analysis, design, implementation and testing of the software, in distinct, non-overlapping phases.
2. The **Architecture** would not be beneficial to projects where the *Game Producers*, *Game Designers*, and anyone else responsible for producing the *game design* did not think in terms of a chain of events.
3. The **Architecture** would not be beneficial to projects where the shared *Game data* and the *Abstract data* were well-defined.

Probably just as important as the items listed, in the limitations of the **Architecture**, are those which are not. Note that there is no description of whether the **Host Modules** (e.g. **Game Controllers Host** and **Events Host**) should be local to each other, on the same machine, or on remote machines. There is no mention of whether you should only have one or more of each type of **Host Module**. There is no limit to how many *Records* each **Host Module** may use to hold information it requires. However, there is an explicit reference, in the description of the **Architecture**, for the need to have only one Central Host and one **Game Database**. Finally, there is no description about what tools could be used to create each type of **Host Module**.

It may be difficult to know when these limitations apply to any particular project. For example, how could you tell when a *game design* was complete? Also how could you tell when the *Game data* and *Abstract data* were well-defined?

5.1 COMPLETE GAME DESIGN CRITERION

Being able to identify an incomplete *game design* would be a per-requisite to deciding whether you should use the **Event-Database Architecture**. To do this, you may consider the criteria used in the *classic software production life cycle*. This begins with an analysis of the requirements the software must meet. The document

DOI: 10.1201/9781003502784-5

containing these requirements is called the User Specifications. In Computer Games, the *game design* is the User Specifications. The criterion for determining a complete set of User Specifications is that it should be possible to write the *User Manual* from the Specifications. This is an old criterion, which seems to have become neglected.

This would mean that a *game design* should contain a description of every stage, in the flow of the game. From the Copyright screens that would be shown when the game starts, to the Congratulation screens that would be shown when the player completes the game, there should be a clear vision of what each stage would be like. A stage could be either a screen or menu or a sub-section of the *Game World*, where a sub-plot would take place. It should be possible to choose any of these stages and give a description to a player of what it would look like, sound like and what options would be available. These descriptions could be at whatever level of detail, short of a breakdown of the methods that would be used to implement each stage (i.e. a *technical design*).

A complete *game design* should contain precise figures for the number of items in each stage: for scores, times, distances and sizes. It should contain a precise description of the position of items on each screen, or menu, or section of the *Game World*. From the smallest item to the largest, all should be described. It should clearly outline whether it would be possible for an item to move, how the item would move, why the item would move, and where the limits would be to its areas of movement. It should clearly describe what sounds each item makes, when these sounds start and stop, whether a sound occurred intermittently or not and how the volume of the sound varies. It should include precise figures for the times between the start and end of a sound, the times between an intermittent sound, the volumes of each sound and the priority of one sound over another. It should also describe any music that could be played during that stage of the game in a similar fashion.

The structure of the document should match the progression of the game. It should begin with the background to the game, and the overall goal. It should then follow the course of the game, beginning with the first stage and ending with the last.

Since the structure of the document follows the course of the game, it should not repeat itself. If two stages of the game shared some of the items of the *User Interface*, then the description of the second stage, in the *game design*, should simply refer back. It should simply refer back to the similar items in the preceding stage. It should not copy these descriptions. If the stages shared, for example, the same options on a menu, commands, characters, creatures or other items in the *Game World*, the *game design* should be structured well enough. So that these could be referred back to, in subsequent chapters, after these had been introduced. Otherwise, the *game design* should not repeat itself.

Instead, a follow-up document, such as an appendix or a *technical design*, should be used to repeat items mentioned in the *game design*. These should group common features of the *game design* and describe each one.

For example, describing all the graphics, sounds or music, which would be used in the game, would give a useful cross-section of the *game design* for a *Game Artist*, a *Sound Designer* or a *Game Programmer*. They could use this to set out their plans and schedules, for creating and using these graphics or sounds.

Another example would be a description of all the computer-controlled characters in a game. This would help a *Programmer* see the common behaviour between these characters. So that he or she could write a *software module*, which could be reused by other *modules* to build of these characters and their Artificial Intelligence.

5.2 INCOMPLETE GAME DESIGN CRITERION

Contrast the complete *game design* just described with an incomplete one. An incomplete *game design* would avoid specifying precise figures for each stage of the game. It would content itself with adjectives like 'little', 'large', 'more', 'less', 'big' and 'small'. It would even turn adjectives into nouns as it struggles to describe its vagueness. It would include words like 'shootable' for an item in the *Game World* that can be shot at, 'moveable' for an item that can be moved and 'jumpable' for a gap in the *Game World* that can be jumped. It would include phrases like 'first playable' to refer to the first version of the game which gave a practical demonstration of the initial stages of its *User Interface*. The most significant of these would be a word like 'deliverable'. This would be used to refer to each version of the game that would be examined, at regular intervals, along the *Software Evolution Process*, by the financial backers of a project or by its leadership. These examinations would mark the points at which the process was meant to be scrutinised. And these would also mark the points at which the feedback was meant nominally to be sought from the software user.

But, of course, the process itself would be unscrutable. So, instead, each examination would be limited to a tactless analysis of the state of the latest version of the product. This would involve either a superficial examination for any obvious defects in each version. Or this would involve an examination of the aggregates and depositions taken from the process so far. These aggregates would include a count of the number of visible errors, how many features were added to the product since the last examination, how many were successfully implemented and the cost of each one in terms of resources, such as time, money or staff. The depositions would not come from the final software user. Instead it would come from the impressions of *Game Testers*, *Game Producers*, financial backers, Creative Directors, Technical Directors and other people whose opinion was considered key who are sometimes called 'Stake Holders' or 'Key Stake Holders'.

One example of how these aggregates would be collected would be through the introduction of time sheets, at the beginning of the production process. These would be used to monitor how much time each member of staff had spent on a task, and how many were involved in each task. Another example would be requests for abortive schedules, from the staff, at the beginning of production. That is to say, they would be asked for schedules which had a purely cosmetic function. None of the times for the tasks would be produced after any analysis had been carried out. And there would be arbitrary limits placed on the time spent on each task. Each could be no longer than two or three days. And anything longer than a week would have to be broken down further.

These schedules would be nothing short of an audacious attempt to regulate a process which has defied regulation since its conception. At the beginning of the *Software Evolution Process*, there would be no complete plan for building the software. There would be no complete *game design* or subsequent documents explaining the plan. So there would be no complete description of the steps which were going to be part of that plan. Consequently, there would be no complete measure of the time each step would take. And there would be no measure of the overall time the project should take. Hence, some other means would be necessary to measure its progress. One of the common alternatives would be to forcibly regiment the remaining steps into the rest of the time allocated for the project, at regular intervals.

Hence, the description of initial set of tasks, given to the staff to produce a schedule from, would be in a vague, abbreviated form, such as items on a menu or a bill. The majority of the tasks would be described in one or two keywords, while a handful would be described in a single sentence. The set of tasks for each member of staff would be assigned to them, and in no particular order. These would not be chosen by them, in response to some problem which they had investigated. There would be no risk attached to each task, by the staff, as a result of this investigation, which would determine the order in which these were performed. Furthermore, the staff would only be given the opportunity to reveal the remaining steps. By breaking any task which they expected to take more than a week, into smaller tasks over two or three days, they would reveal the remaining steps. But they would not be given the opportunity to set the times for these tasks. That would be determined by the forced regimentation of the remaining steps at regular intervals.

Meanwhile, the depositions used to examine a *Software Evolution Process* would be in the form of statements taken from informal interviews. These interviews would include only a small subset of the staff: comprising the leadership and some senior members. After which, the depositions would be combined with the aggregates that had been gathered. And a comparison would be made with another, equally tactless, case study of some other *Software Evolution Process* that had occurred in the past in projects either at the same Software Developer or in another company.

Furthermore, after the examination, the feedback for the process would not directly come from the end-users of the product. Instead, it would either come back, more frequently, from the leadership of the staff. Or it would come back, less frequently, from their financial backers. Or, at the beginning of the process, if a previous version of the game had been released, it would come back from the review of that release, by the Media that cover the Computer Games industry. Or, penultimately, towards the end of the process, it would come back again from any contacts the *Software Developer* had in the Media. Either way, the feedback would be based on some implicit understanding of what the users want. And this implicit understanding would be reflected by the huge gaps within an incomplete *game design*, especially its *User Interface.*

Although the *game design* would include a description of how some parts of the *User Interface* would look, it would not be comprehensive. It would, for example, include a description of a menu that would appear. But it would omit some of the options on that menu. It would omit the position of each option, or it would generalise these positions. Similarly, the *game design* would include a description of a location in the *Game World*. But it would either omit the number of characters, buildings, other structures or other items in that location. Or it would generalise these positions.

Rather than facilitating the writing of a *User Manual*, an incomplete *game design* would read like a *User Manual*. It would use general rules to describe each screen, menu, or section of the *Game World*. But only one instance of each screen, menu or section of the *Game World*, would be described in detail. And this would be repeated throughout the *game design*, for large sets of screens, menus and sections of the *Game World*. The *game design* would use slang terms, from whatever activity the game was trying to authentically capture, without explanation. It would use acronyms, abbreviations and the jargon of the Computer Games industry without explanation.

There would be two signs every incomplete *game design* would exhibit and could not hide. These signs would be clear evidence of its unfinished status. The first sign would be retroactive thinking, which would cause lots of repetition.

An example of this would be the name of an item being mentioned in one chapter, which would only be explained in a later chapter. Another example would be when, in the middle of a chapter describing one topic, the document suddenly interjects with a detailed description of a second topic. But this second topic would belong in a different chapter or would already be described in another chapter. This interjection would be caused by a partial review of the document and a sudden realisation of the vague aspects within it. Instead of following this up with a comprehensive review, a crude attempt would have been made to hastily patch the vague aspects that were found.

The second sign of an incomplete *game design* would be the most damning indictment of all. These would be where the document itself claims to be unfinished. An example of this would be clauses that describe features which would be subject to change. Another example would be statements which referred to features that would be included in future drafts.

But, of course, the moment the rest of the production process commenced, the damage would have been done! And the *Software Evolution Process* would have begun. Any features which were labeled as subject to change, will never be finalized. An incomplete *game design* would remain in a draft state until, and even after, the process has been finished. The reason often given for leaving it unfinished would be that, as the software production progressed, the obscure parts of the *game design* would become clear. The hope would also be that these would not involve any major changes.

Yet this would never be the case. Other parts of the *game design* would be redrafted as well. Different members of the staff would take advantage of the draft state of the *game design*, to make both formal and informal changes. Thus, completing the unfinished parts of the *game design* would never be trivial. The cost, both in terms of finance and time, would spiral upwards.

The signs of an incomplete *game design* would also be evident in subsequent documents. The *software design*, written in the second phase of a *software production life cycle*, would depend on the User Specifications. Similarly, in the Computer Games industry, the *technical design* would depend on the *game design*. The *technical design* would at least refer to, if not completely reiterate, items in the *game design*. It would link these items with a breakdown of the software components, the tools and the methods, which would be used to build these items. The *technical design* would also be used to measure the progress of the production process, and to test the game at the end of it. The finer granularity of its description would make it better suited for these purposes than a *game design*. But only because the game design is so vague and incomplete. In the classic software production life cycle the game design (i.e. User Specification) would be good enough the draw a User manual and a test plan from. In the Software Evolution Process, the game design is not good enough for that.

However, if the *game design* were not complete, the *technical design* would also not be complete. All of its advantages would be lost. The production process would lack objective criteria with which to measure progress. Instead, it would proceed through a lot of trial, and even more errors. This would be another characteristic of an incomplete *game design*.

The software production process would proceed through subjective criteria such as trust and perception. Since the documentation would not be complete, no one would know what the effect of each proposed change could be. So each proposition has to be accepted based on the trust invested in whoever proposed it. And the assessment of the effect has to be based on the immediate perception of the software, by other members of staff.

Thus, the production process would become a melodrama, with the staff as the audience, in which the audience swings from one extreme emotion to next. One moment, the staff would be gripped by anxiety, when a change causes sudden, inexplicable errors. The next moment, they would all be breathing a huge sigh of relief, when someone somehow manages to fix these errors. One moment, a villain emerges for proposing or conducting changes whose immediate effects appear to be catastrophic. The next moment, a hero emerges to save the day, with corrections which would be just as mysterious as the changes that produced the crisis in the first place.

Each scene would draw out personalities, amongst the staff, seeking to play a key role. And they would achieve this ambition either by appearing conscientious or drawing attention to errors caused by other members of staff, in order to gain trust. Or they would manufacture a crisis, if they did not have the patience to wait for one to emerge. By literally rolling their eyes, banging their hands and heads on tables, moaning, groaning or swearing out loud, they would turn any common errors they encounter into a crisis. With loud cries of 'This is bad! This is really, really bad!', 'That is shocking!', 'Oh my God! Oh my God!', and other such exaggerated posturing, they would draw attention. So that they could then suggest ways out of the crisis and improve their perception amongst the staff.

Hence, the melodrama of a *Software Evolution Process* would not produce good personalities, who would seek to complete the *game design*. Instead, it would produce devious personalities who would target those, involved in the process, with the ultimate trust or perception.

In the Computer Games industry, this ultimate trust or perception would be shared by both the leadership and the financial backers of the project. Yet despite the strategic importance of a *game design* to the project, both parties would be ambivalent about it. And this ambivalence would fuel constant debate, between them, about what should constitute a *game design*, throughout the production process.

Although they would all recognise the potency of a *game design*, to sell a project, many would ascribe design to products (i.e. other popular Computer Games) on which no resources have been spent designing. That is, no time, no money and no staff has been spent on its design. Others would believe there was no such thing as a complete design. Some would impatiently push the arguments to the extremes, whenever the subject came up, to immediately cut off any further debate. And they would simply argue that a design could not solve everything.

However, this last argument would be no better than the rest. It would be as equally vacuous as its counterargument. That is, it would be impossible to make a complex product without any design. Both arguments would shed no light on the role of design in software production.

Clearly, the answer would lie somewhere in between the two extremes. It would lie somewhere a million miles to the left of trying to solve everything, and a couple of yards to the right of attempting to build a complex product without any planning. A *game design* (and its *technical design*) would not have to cover all the problems of engineering. Nor would it have to cover all the problems of engineering computer software. Nor would it even have to cover all the problems of engineering the software of Computer Games. It would only need to describe one game.

Many complete designs have been written, in other industries, which could help resolve the debate. The Computer Games industry has not been alone in identifying when a *game design* (and hence a *technical design*) has been complete. Other industries have faced the same problem of deciding when the analyses of the requirements of a product, and the design which will meet those requirements, were complete.

These industries have followed a similar production process to those of Computer Games. This began with an analysis of the requirements of the product. This was followed by a design of the components, and the process used to build and assemble these to meet the requirements. The design was then implemented and the result was tested. These products have been as complex as, if not more complex than, Computer Games. The manufacturers had to produce tangible, durable goods with practical benefits. They did not have the luxury of making products with no practical application. They had to produce goods which, to the consumer, solved challenges, instead of creating imaginary ones. And the production process involved far more disciplines than you would find in the Computer Games industry. The complete designs they have produced are evidence that it is possible in the Computer Games industry, but for a lack of commitment.

In industries, such as the Construction or the Electronics industries, a complete design has been a per-requisite to any project. The *expenditure on design*[1] has taken up a major part of a company's revenue. A major share of the personnel has been committed to ensuring each design was complete.

In contrast, in the Computer Games industry, the initial *game design* has been written by a small group of *Game Designers*. Typically, in a team with 10 *Game Programmers*, there would be 20 *Game Artists*, 1 *Sound Designer*, 1 *Game Producer*, 3 *Game Testers* and 2 *Game Designers*. The ratio of the staff that would be *Game Programmers* would be roughly one-third, while the *Game Artists* would be about two-thirds. The numbers of the other staff, including the *Game Designers*, would not be related to the size of the team. And they would be no more than 3 or 4, in total, no matter whether the size of the team was 40 or 140. Indeed, some *Software Developers* have made games with no *Game Designers*. And their roles have been completely subsumed by other staff, such as the *Game Producers*.

But in most cases, with the exception of the *Game Producers*, and *Sound Designers*, the *Game Designers* have been the smallest contingent involved in the production of Computer Games. And their efforts have to be supplemented by the *Game Producers*, the *Game Artists*, the *Game Testers*, the *Sound Designers* and the *Game Programmers*, after the implementation of the *technical design* has begun. Even though their decisions had the same strategic importance as the *Game*

Producers', the expenditure on the *Game Designers* has not been commensurate. And they have been treated as second-class citizens. This has been a direct result of the little credence given to the *game design*, in the *Software Evolution Process*.

The initial *technical design* too has been drawn up by an equally small subset of the *Game Programmers*. And their efforts have to be subsequently supplemented by the rest of the *Programmers*.

In the Construction industry, companies have been set up which just consist of architects. And the whole business of these companies has been committed to the complete design of a building prior to its construction. Nobody would commission a house or an office with a room or a door missing. But in the Computer games industry it is common place to commission a game with a menu, a level or part of the Game World completely missing.

In the Electronics industry, companies have been set up which produce different parts for computer hardware. To do business, some have completely designed their product and outsourced the manufacturing to other companies. While some have sold the licences for their designs. So that other manufacturers could use it to build and sell the product, as if it were their own.

Both the Construction and the Electronics industries have employed teams of *Drafters*.[2] These employees draw up detailed designs for the buildings and electronic circuits. This has been something they have trained for and specialise in. They have used *Computer Aided Design*[3] (or CAD) software tools to do their work. These tools have been sold on the basis that it was possible to produce a complete design. Derivatives of these tools have, in fact, been used in the Computer Games industry, by *Game Artists*. For example, a popular tool in the Computer Games industry is called 3DS Max by Autodesk. Autodesk also made AutoCAD which 3DS Max is derived from. So there has been clear evidence of the viability and importance of a complete design to a production process. Therefore, the scarcity of complete designs, in the Computer Games industry, does present a major challenge.

Some in the industry have intuitively believed the greater complexity of the production process of Computer Games has been responsible for scarcity of complete designs. However, if you were to compare the two classes of industry which the Construction, the Electronics and the Computer Games industry belong to, you would find that this could not be the case.

The Computer Games industry belongs to the Software industry. The Software industry has been a service industry. The products the industry has produced partially, or completely, automated the existing processes of their clients. Thus, they could see a working model which their products would substitute.

Whereas the Construction and the Electronics industry have been manufacturing industries. Their clients did not have a working model for which they would provide a substitute. So the production processes of these industries had to produce a simulation of a working model, before producing the product itself. This has meant there have been more component parts of these processes, and more dependencies between these components, than in the service industries.

The Computer Games industry has been a service industry. And their processes have been no more complex than the rest of the Software industry. Within any given year, a few *original games* may be released, which have no existing working model to help the production of these games. But the vast majority would not be original. These would either be based on films, books, well-established sports or other forms of games, such as card games and board games. Or these would be based on previously successful Computer Games. The companies which produced these games would have neither the inclination, let alone the resources, to carefully construct, for example, an original novel. Before subsequently building a game based on that novel. If they did, the time it would take to produce these games would be twice as long. Instead of a period of between 18 and 24 months to produce each game, it would take around three to four years.

Even assuming all of the games were original, the *Software Developer* would have been in an even more advantageous position than a company in the manufacturing industries. Whereas in a manufacturing industry, like Construction, one company may be commissioned to produce a design for another to use. In the Computer Games industry, the client would be relying on the *Software Developer*, to produce the designs which they were going to use. So the *Developer* would have the opportunity to set the requirements which they themselves would meet.

The client may interfere with the designs of the *Developer*. Or the client may interfere with the production process, by demanding to periodically see the state of the unfinished product. Some would like to believe that this makes the production process of the Computer Games industry more complex than the others.

But, like the Construction and the Electronics industry, the clients want the product to meet their requirements: not the *Developers*. Like these other industries, the clients want to be able to scrutinise the production process. Perhaps the clients have more intelligence than the Computer Games industry gives them credit. Perhaps the clients recognise that some of the definitions of a complete design and a production process, in the industry, lacks credibility. But the clients have not been any more demanding than those of the other industries.

Some believe that the need to save material resources has been the reason why a complete design has been so important to manufacturing industries. Since there were no expensive materials involved in making changes to a product, in the Computer Games industry, a complete design has had less significance. But there would be more advantages to having a complete design than saving material resources. These would be, namely, saving time, money and ensuring the quality of the product.

Saving time saves money, no matter what your industry. Even if you did intend to revise a game several times, the advantages of being able to understand the effects, of your changes, would outweigh the time lost writing up a design. In fact, there would be no limit to how much time could be lost due to a lack of understanding. Nor could anyone measure the cost of a loss of reputation, when these changes caused errors which surfaced later on, after the product had been released. Cutting back on quality would lose you money, no matter what industry you were in.

The Computer Games industry has catered for quality, but only for *Quality Control*[4] not *Quality Assurance*.[5] Not withstanding that the third phase of the production process of Computer Games being called 'QA', and the presence of departments in the Software Developers called 'QA Departments', the Computer Gamed industry has never provided Quality Assurance. It has only provided Quality Control. This has occurred when, just prior to its release, a product has been tested against its requirements. This process, however, has already been undermined by the fact that these requirements i.e. the *game design* was not completed. The testing process has been more empirical than methodical. The process has been further undermined by the fact that the tests were carried out against the *game design*, and not the more in-depth *technical design*.

As a result, the testing of the software has not been exhaustive. Crude substitutes have been deployed, instead, to perform some kind of exhaustive test. This has included *Soak Testing*.[6] Normally, this form of testing would be used to identify problems with the performance of a system, after it was complete, and there were no errors in it. This would test the system under high usage over an extended period of time. But, in the Computer Games industry, these tests have been used to catch errors in software, before it was complete. Furthermore, this has been abused to give the production process credibility: by implying the software was complete.

However, even *Soak Testing* requires a complete analysis of the components of the system to be tested (i.e. a complete *game design* and *technical design*). And in the absence of this analysis, these tests have no meaning and remain empirical. Thus, despite the introduction of jargon such as *Soak Testing*, into the process of testing Computer Games over the years, to give it credibility, empiricism has continued to dominate it. That is, the staff have been left to assess the product based on their own senses, personal instincts, experiences, emotions and intuitions. Although a few of them do, none of them have had to rationalise why some aspect of the software constitutes an error. And the few that do would not base their rationalisation on an agreed set of facts (such as items in a *game design*).

Along with the credence given to empiricism in the *Software Evolution Process*, a lot of credence has been given to self-assessment to create a standard of quality for the final product. That is to say, a lot of trust has been placed in different members of staff testing the quality of their own work. Sometimes, they vary this scheme, by requiring staff to have their immediate colleagues review and approve their work before it is submitted into a Software Repository or archive of files used to build the game. They call this scheme, Buddy Checking. But it amounts to the same thing. And that, somehow, this would naturally produce a consistent, high standard of quality throughout the final product.

Intuitively, most industries would expect staff to be more lenient when judging their own work than when judging others. And that self-assessment in a collaborative venture had no hope of creating a product of consistent quality. Since different staff would judge their own work by different standards. In the Computer Games industry, however, this notion has become incredulous. Likewise, the notion that a handful of dedicated staff, such as the *Game Testers*, had no hope of being used to set up a consistent standard for assessing quality has also become incredulous.

It is in keeping with industry's spirit of empiricism and self-assessment that the *Game Testers* have been left to their own devices: to use their own experiences and intuitions to assess the final product. Most of them have been temporary, seasonal workers, who only joined the production process at the end of it. Most look at the position as a merely temporary one. They have used it, and it would be presented them during their interviews, as a stepping stone to achieve a more senior position, as a *Game Artist*, *Game Programmer* or even a *Game Producer*. The only qualifications they required were a passion for Computer Games, to have spent an unhealthy amount of time playing old games, and to be willing to work many more hours, days, months or years testing new ones. They have taken little or no part in the drafting of the initial *game design*. In short, they have been treated like second-class citizens. And, just like the *Game Designers*, this has been a direct result of the little credence given to the *game design*, in the *Software Evolution Process*.

Their testing process has been limited to the view of the quality of the final product i.e. Quality Control. It has completely ignored the production process. If any decisions were made which wasted a lot of time, or caused major problems later on, these have been swept under the carpet. Quick and dirty solutions have been used to fix the errors that had arisen as a result. These have been the only solutions possible. During the final stages of a long production process, there is no way you can turn a low-quality product into a high one.

Compare this approach with the Construction or the Electronics industry. These industries have understood that quality is a process, not a product. This has been argu-ably why they have paid so much attention to their designs. They have realised that through a design, you have a chance to carefully examine your ideas and, more impor-tantly, the way you think. The way you think determines the way you work. The way you work controls what you produce. Thus, by ensuring their minds, they have ensured their products. Or as the proverb says, *watch your character, it becomes your destiny.*[7]

The word *quality*[8] simply means characteristic. When it has been applied to a product, it has meant that the characteristics of the product met some, or all, of a con-sumer's needs. No person or product would achieve a characteristic by chance, but by habit. A product must always behave a certain way for that behaviour to be a char-acteristic. No human-made product would behave in this manner except by design. Therefore, it would be impossible to achieve *quality* without design. This would apply to each individual characteristic that makes up the *quality* of the product.

Without a design, there would be no link between a consumer's needs and a pro-ducer's service: between a problem and a solution. When *quality* was applied to a prod-uct, without a design, there would be only one perspective left to look at it from. This would be, namely, the producer's needs and requirements. These needs would only be coincidentally related to a consumer's needs. The producer's need to sell or improve their products or services would only be possible if the consumer happened to need, and were willing to pay for, the *quality* of the current product or service. And, eventu-ally, the producer would subsequently find out that what they thought was high *quality* had been ignored in the market place. These rude awakenings have been why so many, the Computer Games industry included, have found *quality* so hard to describe.

The fact that a design relates a problem to a solution has also been why they have found design so hard to describe. They have either unconsciously forgotten the prob-lem. Or they consciously ignored the problem because they have been afraid. They have been afraid of looking at a problem in detail, and the long list of solutions they would have to commit to as a result. Instead, they have sought the licence to be free of a consumer's needs, or the constraints of a problem.

Once, of course, they had freed themselves from the constraints of themselves from the constraints of the problem, it has then been very difficult to define what design is. And it has been very easy to believe that any implementation (especially any novel solution) was synonymous with a design. This has been exactly what has happened to the Computer Games industry. In other industries, such as Construction and Electronics, the *role of design*[9] has clearly been defined by the need to explain the production process. In the Computer Games industry, the role of design has been something mysterious, which has only physically manifested itself in novel solutions.

When some novel (or complex) feature has been implemented, it has been assumed that this feature has been designed. When a popular or successful Computer Game has a novel or complex feature within it, it have assumed that that feature and by extension the game has been designed. Someone must have thought about it, before adding it. Therefore, it must have been designed. It has not mattered that the person, who did this, did not write it down in any comprehensible manner. Nor has it mattered that neither that person, nor anyone else, could repeat the same creative process.

Of course, the assumption of a causal relationship, between the implementation of a novel (or complex) solution and a design, has been false. You could implement such solutions by experimentation, duplicating features from other products or combining both these techniques. You could keep adding different features to a game, piece-meal, like a sculptor or a painter, till serendipity offered you a crude gift.

Of course, like a piece of sculptor, you could not reproduce exactly the same result, using the same tools. Nor could you identify and examine the component parts of a statue, or the process that made it. Instead, these three features would be the characteristics of a true design, a complete *game design* and *technical design*, and its advantages over a false one:

1. You could follow the design to produce the same product every time.
2. You could identify and test the component parts of the product.
3. You could identify and test the component parts of the production process.

Since you could identify the component parts of a product which had been designed, you could simplify the testing process. You could test the different parts individually, in a modular form. This would reduce the complexity of always testing the entire product, along with many interdependent parts, to a simple series of smaller tests, of independent parts.

So when you began testing a product, whose requirements were analysed, and manufacture was designed, you should know how many features would be tested. This should be clear from the component parts, and the features in each. The total number of untested features should begin from a fixed number and go downwards. It should never rise above its start point. For a product whose requirements were not analysed, and manufacture was therefore not designed, the number of untested features would be arbitrary to begin with. What is more, it would rise above its start point.

In the Computer Games industry, the number of untested features would typically be stored in a *Database*. This *Database* would be controlled by software known as a *Bug Database*. After the game had been released, this would normally be used to produce charts and graphs, showing the progress of the total number of untested features, and errors, from the beginning to the end of the testing process. These graphs would provide clear evidence of an incomplete *game design*.

Some *Software Developers* introduce this *Bug Database* at the beginning of the production process. And the reference to this tool, in the *game design* or *technical design*, at this point of a process would also be evidence of an incomplete *game design*. For the gaps in the *game design* would be logged as errors in the *Bug Database*. And the *Database* would be abused as a device for designing the product, as supposed to testing it. This would happen due to the absence of any clear plan for building that product. If there were a plan, and this were recorded in some documentation, such as a *game design* or *technical design*, there would be no need to keep another record of what items were missing, in a *Bug Database*. Since the document itself would make it self-evident, what items were missing in the product?

In fact, the *classic software production life cycle* uses this criterion to identify a complete *software design*. That is to say, a *software design* would be complete if you could produce a test plan from it. In the Computer Games industry, the *technical design* is the *software design*. So if it were not possible to produce a clear plan, of how a game would be tested, from its *technical design*, then that would indicate that its *technical design* was not complete. And by implication, this would provide further evidence that its *game design* was not complete.

By itself though, this evidence would not be conclusive. The *game design* could have been completed, but the *technical design* was not. However, this would be conclusive evidence if it were accompanied by either the fact that every component, of the *game design*, had been covered in the *technical design*. Or, if it were accompanied by the other signs of an incomplete *game design*, that have already been mentioned.

5.3   COMPLETE DATA DESIGN CRITERION

Being able to recognise when the *Game data* and the *Abstract data* were not well-defined would also be a requirement, before deciding to use the **Event-Database Architecture**. These *data* would normally be described in the *technical design* for a project. So you could use the same criteria used to detect a complete *technical design* to detect when these *data* were well-defined. This criterion is, as has already been mentioned, whether it is possible to produce a test plan from the *technical design*.

For the *Game data* and the *Abstract data*, this means it should be possible to provide a clear plan for modifying each one. And the plan should include a method for monitoring the effects of each modification. Furthermore, the results of each modification should be predicted in the plan.

For example, suppose some *data* would be used to display an item, lying around in one part of the game. It should be possible to modify that *data*. And it should be possible to see the change in the appearance of that item, in the game.

Another example, suppose some *data* would be used to play a sound, when a player performs one command. It should be possible to modify that sound. And it should be possible to hear the new sound for that command.

Another example, suppose some *data* would control the speed of one of the characters that moved around in the game. It should be possible to modify that speed. And it should be possible to measure the new speed of that character.

For each piece of *data*, it should be possible to know its limits. That is to say, the range of values, the different formats, the minimum and maximum size, or the maximum length of that *data* should be clear. And it should be possible to give that *data* an invalid value, an invalid format, an invalid size or an invalid length and observe the effects. The effects of these errors should be accommodated in the definition of the *data*. So that this definition could be compared with the effects of the errors manufactured during the test.

Another criterion that may be used, to detect when the *Game data* and *Abstract data* were well-defined, has already been mentioned in the description of the **Game Database**. The **Game Database**, of an **Event-Database Architecture**, too would be required to be well-defined. This would be determined by its transparency to all the members of the staff involved in the production process.

That is to say, it should be possible for the members, using a variety of software tools or hypothetical speculations, to modify the *data* effectively to achieve a goal. If the definitions of the *data* were completed, each use of that *data* should be consistent with another. Each person's understanding of that *data* should, at least, be consistent. This should apply to the general *Game data* and the *Abstract data*.

Some *Abstract data* may require extensive knowledge of an academic field (e.g. Mathematics). So, in these cases, only those with that knowledge should be able to modify these *data* to effect. However, anyone, without knowledge in that field, should still know how these *data* were related to the general *Game data*. And that this relationship was an academic field (e.g. Mathematics). And, in the case of the Event-Database Architecture, the Entity-Relationship diagrams accompanying the data design should also show this relationship. Since all specialised *Abstract data* would be derivatives of the general *Game data*. These *data* would be a combination of general *Game data* and *data* from a specialised field. So the *Abstract data* should still use the language from the description of the general *Game data*. And hence, everyone should still be able to identify the relationship of any *Abstract data* to the *Game data*.

5.4   INCOMPLETE DATA DESIGN CRITERION

Some of the symptoms, in a project, of the *Game data* and the *Abstract data* not being well-defined have already been described. These were included in the description of the problem that the **Event-Database Architecture** attempts to solve (see the chapter entitled **The Problem**).

Briefly, these included the inconsistencies in the quality of the descriptions of the *data*. These also included the selection of arbitrary names and words, by the members of staff, to describe that data, based on their background and experience. Finally, these included the incorporation of these arbitrary names and words into a degenerative language for communicating between the staff. This has a degenerative effect on their productivity. And that, in turn, affects the overall success of the project, and its ability to innovate.

However, there are further symptoms of poorly defined *Game data* and *Abstract data*. These relate to how a game would be tested at the end of a software production process. When the *data* has not been well-defined, it would not be possible to test that *data*.

It would not be possible to predict what the effects of modifying that *data* would be. So you could not plan any comprehensive test of that *data*. Either some of the *data* would have no description at all. Or some would only have a partial description. So you could not predict what the effects of all the possible modifications of that *data* would be. Or some descriptions would not include the limits of that *data*, and an explanation of what would happen if these limits were exceeded. That is to say, the descriptions would not include the range of values, the different formats, the minimum and maximum size, or the maximum length that *data* could take. So you could not predict what to expect if you attempted to modify that *data* erroneously.

The ultimate symptom of poorly defined *Game data* and *Abstract data* is the lack of transparency of these, to the members of staff. Either all the members cannot tell, only from the description available of any given *data*, whether that *data* was general *Game data* or specialised *Abstract data*. Or, when they can distinguish between the two classes, they cannot relate that data in one class, to some data in the other class. So they cannot understand, from the description of any given *Abstract data*, which *Game data* it is related to and how. Nor could they understand from the description of any given *Game data*, which *Abstract data* it was related to and how.

NOTES

1. *Expenditure on design.* An analysis of the relationship between expenditure on design tools and competitiveness of Integrated Circuit companies, conducted by the Electronic Design Automation Consortium (EDAC), showed that there was a strong link between the investment in design and a company's future market position. See Glossary.
2. *Drafters.* A profession which prepares technical drawings and plans used by production and construction workers. These drawings are used to build everything from manufactured products (e.g. toys) to structures (e.g. an office building). See Glossary.
3. *Computer Aided Design (software).* Computer software used to design and simulate physical tests of products which require expensive raw materials before these are physically manufactured, such as buildings, cars and electronic circuits.
4. *Quality Control.* A system that accepts or rejects products or services depending on whether these meet all of the customer's specifications and requirements. See Glossary.
5. *Quality Assurance.* In theory, a system which ensures that a company's processes (as supposed to their product) will meet all of the customer's requirements and specifications. In practice, software companies just apply two Quality Controls in the latter stages of production, known as Alpha and Beta testing, and call it Quality Assurance. See Glossary.
6. *Soak Testing.* A process for testing a complete software or hardware system, to reveal errors that only emerge under extreme conditions. See Glossary.
7. *Watch your character, it becomes your destiny.* Quotation from Frank Outlaw, American business man and retail executive.
8. *Quality.* The characteristic of a product which meets a customer's needs. See Glossary.
9. *Role of design.* In industries such as Construction and Electronics, the role of design is not just to describe the plan for making a product but also to analyse the interaction of the different parts of the plan. See Glossary.

6 Glossary

6.1 CLASSIC SOFTWARE PRODUCTION LIFE CYCLE

A production process that follows the analysis, design, implementation, testing, installation, maintenance and retirement of software. The process requires extensive documentation; at least six in all. These include a description of the requirements of the software, the preliminary software design, the User Interface, a final software design, a plan for testing the software and a User manual.

Since commercial software, such as computer games, are released after being tested, there is usually no maintenance and retirement phase. The licences that accompany these software usually explicitly deny any warranty and do not claim any suitability of the software for any purpose. Any maintenance done on it is at the user's personal cost and risk.

The software production life cycle is also called the Waterfall. It is one of the oldest software production processes. It borrows a lot from the production processes used in other forms of engineering. These include civil, mechanical and electrical engineering. Many subsequent software production processes are based on it.

There is a common fallacy that the Waterfall is not adept to changes in a customer's requirements. The fallacy comes from the fact that the Waterfall is a linear process i.e. no two phases of it should overlap. The misconception of this fact is that each phase has to be completed before the next begins. This misconception is often deliberately spread to make the Waterfall look very naïve, prior to the presentation of a convoluted and expensive substitute.

It is true that no two phases of the Waterfall should overlap. But if it becomes clear, from an encounter with an unexpected problem, in the current phase, that the previous phase was incomplete, then you simply go back to the previous phase. You keep doing this until you have to return back to the original analysis.

A software life-cycle or product life-cycle model, described by W. W. Royce in 1970, in which development is supposed to proceed linearly through the phases of requirements analysis, design, implementation, testing (validation), integration and maintenance. The Waterfall Model is considered old-fashioned or simplistic by proponents of object-oriented design which often uses the spiral model instead....

Source: Waterfall. The Free On-line Dictionary of Computing
© 1993-2001, Denis Howe

Figure 3 portrays the iterative relationship between successive development phases for this scheme. The ordering of steps is based on the following concept: that as each step progresses and the design is further detailed, there is an iteration with the preceding and succeeding steps but rarely with the more remote steps in the sequence. The virtue of all of this is that as the design proceeds the change process is scoped down to manageable limits. At any point in the design process, after the requirement analysis is completed, there exists a firm and closeup, moving baseline to which to return in

DOI: 10.1201/9781003502784-6

the event of unforeseen design difficulties. What we have is an effective fallback position that tends to maximize the extent of early work that is salvageable and preserved.

…The first rule of managing software development is ruthless enforcement of documentation requirements.

Occasionally I am called upon to review the progress of other software design efforts. My first step is to investigate the state of the documentation. If the documentation is in serious default my first recommendation is simple. Replace project management. Stop all activities not related to documentation. Bring the documentation up to acceptable standards. Management of software is simply impossible without a very high degree of documentation….

Why so much documentation?

1. *Each designer must communicate with interfacing designers, with his management and possibly with the customer. A verbal record is too intangible to provide an adequate basis for an interface or management decision. An acceptable written description forces the designer to take an unequivocal position and provide tangible evidence of completion. It prevents the designer from hiding behind the – "I am 90 percent finished" – syndrome month after month.*
2. *During the early phase of software development, the documentation is the specification and is the design…If the documentation does not yet exist there is as yet no design, only people thinking and talking about the design which is of some value, but not much.*
3. *The real monetary value of good documentation begins downstream in the development process during the testing phase and continues through operations and redesign…*

a) During the testing phase, with good documentation, the manager can concentrate personnel on the mistakes in the program. Without good documentation every mistake, large, or small, is analysed by one man who probably made the mistake in the first place because he is the only man who understands the program area…

c) Following the initial operations, when system improvements are in order, good documentation permits effective redesign, updating and retrofitting in the field. If documentation does not exist, generally the entire existing frame work of operating software must be junked, even for relatively modest changes…'

Source: Managing the development of Large Software Systems © 1970, Institute of Electrical and Electronics Engineers. Dr. Winston W. Royce

6.2   SOFTWARE DESIGN

A breakdown of the software components, tools and techniques that will be used to build and assemble software that meets a User Specification i.e. a customer's requirements. A breakdown of the software procedures and data that will be used to build a software module.

6.3   SOFTWARE MODULE

A small piece of software, which is a component part of a larger computer programme. Each solves one facet of the overall problem the programme was designed for.

6.4 SOFTWARE DATA

Information suitable for computer processing.

6.4.1 GAME DATA

General information which is shared between software modules e.g. the name of items in the Game World, commonly used text, 2D images, 3D models sound and so on.

6.4.2 ABSTRACT DATA

Special information which is designed to be only used by a single software module. In the Computer Games industry, because often there is no complete game design and hence no complete data design, there is a lack of a well-defined data. So although some of the data is called and treated as if it were Abstract Data, almost all data is a form of Game Data which is shared by multiple software modules.

6.5 SOFTWARE LIBRARY

A collection of computer programmes, or software modules, which perform a commonly repeated task on computer hardware e.g. reading data from, and writing to, computer files, rendering graphics, playing sounds.

6.6 GAME DESIGN

A term sometimes used to refer to the User Specification of software, in the Computer Games industry. It is a description of the goal of a game, the different stages, the progression and the User Interface through these stages.

6.7 INTERFACE

A common boundary.

6.7.1 USER INTERFACE

The set of components (e.g. images, messages, commands or menu options) that allows a user to interact with software.

6.7.2 PROGRAMMING INTERFACE

The set of components (e.g. procedures or data) that allows one software module to interact with another.

All software that a user can interact with has a User Interface, and at no point, while it is being used, does it not have one. Whether it is displaying 3D or 2D images, or plain text, these are all part of the User Interface. Even a photo-realistic 3D display of the Game World is still part of the User Interface. It is not real. It is merely a representation of the data within the computer system, which you can interact with, whose function is to inform the users, not to fool them.

The aspects of a computer system or program which can be seen (or heard or otherwise perceived) by the human user, and the commands and mechanisms the user uses to control its operation and input data.

A graphical user interface emphasises the use of pictures for output and a pointing device such as a mouse for input and control whereas a command line interface requires the user to type textual commands and input at a keyboard and produces a single stream of text as output.

Source: User Interface. The Free On-line Dictionary of Computing
© 1993-2001, Denis Howe

6.8   TECHNICAL DESIGN

A term sometimes used to refer to a software design, in the Computer Games industry. It is a description of the software modules, data, tools and techniques that will be used to implement a game design.

6.9   GAME MODULE

A software module which is used to implement unique aspects of a game.

6.10   GAME ENGINE

A set of software modules (or library) which were designed to be reused to make many aspects of different games. Or that is what is meant to happen in theory. In practice, the library originates from a game of a certain genre. And it is only suitable for making other games that fit into that genre of the original game.

6.11   USER MANUAL

An instruction booklet, for software users, which explains how to solve a problem using the software. In computer games, this includes a description of the problem (i.e. the background and goal of the game) and a description of how to use the Interface of the game to solve the problem.

6.12   PRE-PRODUCTION, PRODUCTION AND QA

The differences between the names of the phases, of the production of Computer Games and other software, stem from the crisis of identity which the Computer Games industry suffers from. The industry has a hard time deciding whether it is part of the Film industry or the Software industry. While some in the industry liked to think of themselves as Film Studios, others liked to think of themselves as Software Houses. This conflict has subsequently been reflected in the names of the phases of production. The name of the first two phases comes from the affinity with the Film industry, which uses the same names to refer to the initial phases of film production. While the name of the last phase, short for Quality Assurance, comes from the affinity with the Software industry, which uses the same name to refer to the final phase of software production.

Along with the first two phases, there is a fourth phase in the Computer Games industry known as Post-production. This occurs after the software has been released and has nothing directly to do with its production. Its effect is only posthumous and it is sometimes omitted. The name of this phase too comes from the final phase of film production.

Regardless of the size of the team, scope of the game, the budget, or anything else, a basic framework exists for the overall production process. The process can be broken down into four broad phases: pre-production, production, testing, and post-production....

Pre-production is the first phase in the production cycle and is critical to defining what the game is, how long it will take to make, how many people are needed, and how much everything will cost. Pre-production can last anywhere from one week to a year or more, depending on how much time you have to complete the game...

The production phase is when the team can actually begin producing assets and code for the game. In most cases, the line between pre-production and production is fuzzy, as you will be able to start production on some features while some features will still be in pre-production...

Testing is a critical phase in game development. This is when the game gets checked to ensure that everything works correctly and that there are no crash bugs. Testing is ongoing during the production process, as the Quality Assurance (QA) department will check milestone builds, new functionality, and new assets as they become available in the game...

After the game is code released and approved for manufacturing, the game development process needs to be wrapped up before it is officially completed. Many times, this step is forgotten or ignored, which is unfortunate...The post-production phase consists of two things: learning from experience and archiving the plan.

Source: Game Production Handbook © 2006,
Charles River Media. Heather Chandler

6.13   SOFTWARE DEVELOPER

A person or company that produces software.

6.14   SOFTWARE ENGINEERING

A systematic, disciplined approach to software production. It was devised to cope with large projects which no one individual could undertake to deliver in a timely, secure fashion.

This approach may begin with a prototype. A prototype is the first product of the software production process. All other products of that process have the same qualities. So the prototype can be used to assess the feasibility of the process.

Software Engineering: (1) The application of a systematic, disciplined, quantifiable approach to the development, operation, and maintenance of software; that is, the application of engineering to software. (2) The study of approaches in (1).

Source: IEEE Standards Collection: Software Engineering, IEEE Standard 610.12-1990 © 1993, Institute of Electrical and Electronics Engineers

Software engineering is the establishment and use of sound engineering principles in order to obtain economically software that is reliable and works efficiently on real machines.

Source: Software Engineering: A Report on a Conference Sponsored by the NATO Science Committee © 1969, North Atlantic Treaty Organisation.
Naur, P. and B. Randall (eds.)

6.15 GAME-EDITORS

One or more tools that allow the elements of a game to be edited. These elements may either be menus, locations, characters or other items that appear in the Game World.

For example, the game-editor may allow you to edit the number of characters roaming around or other items lying around in one location of the game. Or the game-editor may allow you to edit the position of these items in that location. Or these may allow you to edit the shape of that location. Or these may allow you to edit some of the properties of a character or another item in that location. This includes the colour, size, appearance, animation, sounds or pattern of behaviour of that character or item.

Originally, game-editors were only produced at the end of a software production process, sometimes after a game had been released. These were meant to give the end-users or players the ability to either add new features to the Game World or extend existing ones. But later, these editors began to appear at the beginning of the production process. So that these could be used by Software Developers to edit the Game World when the game design changed.

However, unlike the original editors, these later ones had the additional problem of trying to edit a game, with an incomplete game design, that kept changing. If there were anything more difficult than trying to produce a game, with an incomplete game design, it would be trying to produce the software to edit one.

A popular example of one of these modern editors is the Unity Editor, which comes with its own game-engine called the Unity Engine. There are at least five popular points used to market the Unity Editor, and its advantages, to the production of computer games. These are that it enables you to

1. build impressive looking 2D or 3D games on your own
2. rapidly build a small section of a game, or 'prototype' in the language of the Computer Games Industry, which can be used in its feasibility study or to market it

3. change properties, appearance and physics of items in the game, while playing the game, which makes the project seem very flexible
4. extend features of the Editor by writing your own tools with its programming languages, or purchasing software from a large group of third parties,
   which again makes the project seem flexible
5. build games for different or future Operating Systems or Computer
   Hardware, or 'platforms' in the language of the Computer Games Industry,
   which makes the project open to more or new lucrative markets

But all of these advantages do not hold up when examined from the perspective of
a formal definition of Software Engineering (see the definition of software engineering and prototype in the Glossary).

Briefly, Software Engineering is a systematic approach to building software
which requires a lot of people, with different skills, to collaborate. It is a systematic
process developed to ensure the success of large projects. The first product, from
the beginning to the end of this process, is called a prototype, which may be used to
assess the feasibility of that process.

From this definition, it is clear that with respect to the Unity Editor,

1. the Unity Editor tries to bypass any need for collaboration and provides a
   single tool that allows a single person to build a game alone, so no matter
   how impressive the 2D or 3D graphics of the game look, if you use the
   Editor for a large project, that project will fail, according to the requirements of Software Engineering
2. the prototype you produce with the Unity Editor is not the first product of
   a production process, it does not come out of running through the process
   from the beginning to the end, but it is a product of ad hoc experimentation
   at the beginning
3. editing the properties of a game, while playing it, is an ad hoc process and
   not a systematic one required by Software Engineering
4. there is no systematic way with the Unity Editor to describe some problem with a Software Design, Architecture or tool used in a production
   process and find a solution for it either by writing your own extension
   to the Editor through its programming languages or buying third-party
   tools
5. the Unity Editor and Unity Engine are not scalable

In addition to (2), it is worth noting the following about the prototype produced
with the Unity Editor:

1. the prototype does not have the same qualities as the rest of the products of
   the production process but is merely a small section of the Game World or
   'Unity Scene' in the language of the Unity Editor
2. the prototype will probably not even be in the final product
3. the prototype cannot be used to assess the feasibility of the production
   process

In addition to (3), it is worth noting the following about editing a game while playing it in the Unity Editor:

1. the changes depend on incidental causes (i.e. observations that happen to occur while playing the game)
2. the changes depend on incidental effects (i.e. adjusting the properties of items or game modules, in the Game World through a process of trial and error to correct problems or unwanted behaviour in the observations)

In addition to (4), it is worth noting the following about what the Unity Editor provides you with:

1. a tool (i.e. the Unity Asset Store) which allows you to search for third-party tools that can solve a problem
2. searching the Asset Store depends on Keywords for different areas of the game (e.g. '3D Models', 'Animations', 'Audio', 'scripting' etc.) which are just too vague for a systematic approach and requires a process of trial and error to find the right tool

In addition to (5), it is worth noting the following about the Unity Editor and Unity Engine:

1. there is no systematic approach to transferring a game from one 'platform' to another
2. the Software Developer has to be aware of which 'platform' the Unity Engine is working on and select the right features to use, when they write the software
3. the Unity Engine does not have a paradigm or model or way of working, which works across all 'platforms' which is required for it to be scalable
4. there is no systematic approach for even transferring the game from working within the Unity Editor to working without the Editor, and there are features which the Developer has to be aware will only work in the Editor and cannot be used to build the final product without the Editor
5. the Unity Editor and Unity Engine were both produced by an ad hoc approach (i.e. a Software Evolution Process) and not by a systematic approach (i.e. Software Engineering) which would be required to produce scalable software, as a result amongst other things the documentation of the Editor is poor, features are continuously being added or removed,
6. the Unity Editor is currently (in 2015) only available for 2 'platforms' (i.e. Microsoft Windows and MAC OS X)
7. the Unity Engine is available for over 20 'platforms'
8. if the Developers of the Unity Editor and Unity Engine could produce scalable software, why is it that the number of 'platforms' for the Editor and Engine differ so greatly?
9. if the Developers of the Unity Editor and Unity Engine could not produce scalable software themselves, then how can they provide others with tools that do produce scalable software?

Nevertheless, the Unity Editor is very popular, and it has many admirers. Some would point to this fact, namely there being many articles and forums on the Internet, some from the Developers of the Editor and Unity Engine, and others from fans who give advice. And you can search for advice on how to transfer a game from one 'platform' to another. And this search is part of the paradigm or model or way of working of the Unity Editor. Hence, the Unity Editor and Unity Engine have a single paradigm for all 'platforms' and therefore are scalable.

This searching process may or may not be part of the paradigm of the Unity Editor. But, nevertheless, this search process is ad hoc. And in no way shape or form can an ad hoc process be described as scalable, let alone Software Engineering.

Unity is a flexible and powerful development platform for creating multiplatform 3D and 2D games and interactive experiences. It's a complete ecosystem for anyone who aims to build a business on creating high-end content and connecting to their most loyal and enthusiastic players and customers.

Source: The best development platform for creating games
(c) 2015. Unity Technologies

Most of Unity's API and project structure is identical for all supported platforms and in some cases a project can simply be rebuilt to run on different devices. However, fundamental differences in the hardware and deployment methods mean that some parts of a project may not port between platforms without change. Below are details of some common cross-platform issues and suggestions for solving them.

Source: Porting a Project Between Platforms (c) 2015. Unity Technologies

6.16 PLATFORM

A marketing term for a computer hardware or Operating System or third-party game-engine that a computer game can be built and sold on.

6.17 SOFTWARE EVOLUTION PROCESS

A name given to any of a large set of ad hoc, non-linear, software production processes, that are meant to evolve a piece of software to meet a software user's requirements. These are based on rapid feedback from the user.

The concept of a Software Evolution Process was devised by Meir 'Manny' Lehman, while he was working for the International Business Machines Corporation (IBM). He was investigating the level of productivity occurring on the OS/360 Operating System for mainframe computers.

He used the term to describe a phenomenon that occurred to software which was continuously being adapted to meet a changing market. He noticed that these software all rapidly became too complex, and the production process grounded to a halt. The software then had to be dramatically revised or replaced by something new. His contemporaries at IBM, such as Fredrick P. Brooks, who looked at his study, agreed with him. Fredrick P. Brooks went on to write The Mythical Man Month, documenting the phenomenon, which was originally published back in 1975. Lehman believed there was a theory (or set of laws) which explained this phenomenon. And that if you could understand this theory, you could extend or prevent these terminal processes. He called it the theory of Software Evolution.

Over the years, the Software Evolution Process has been re-branded and reintroduced into the Software industry, several times, as a new methodical technique. But these techniques have neither been new nor methodical. Two such examples of this re-branding and reintroduction have come in the form of two methodologies known as Extreme Programming and SCRUM.

Both are a subset of a methodology known as Agile Software Development Management. And Agile Software Development Management is a subject of an academic discipline known as Project Management.

One of the central premises of Project Management is that the development of a plan, for any project (including software production), is an iterative process. That is, the plan will change as time goes by, and you learn more and more about that project. Another central premise is that the risk associated with a project should be kept to a minimum, by breaking it down into smaller tasks. And by accounting for the cost and time to complete each of these tasks.

The effect of the first premise, in software production, is that Project Management creates iterative software production processes, with little or no advanced analysis. The effect of the second premise is that Project Management creates non-linear software production processes. Since by overlapping phases of the classic software production life cycle, the time at least, if not the cost of a project, can be kept to a minimum. And hence the risk can be kept to a minimum.

Therefore, both Extreme Programming and SCRUM bears these two characteristics! These processes are both iterative and non-linear. These two characteristics are also the defining characteristics of a Software Evolution Process.

SCRUM is the most common form of Agile Development used in the Computer Games industry. Since it has fewer rules than Extreme Programming. But both SCRUM and Extreme Programming share a lot of terminology which makes it sound as if these methodologies originated from software engineering even though these originated from Project Management e.g.

'Test-Driven Development',
'Refactoring',
'Continuous Integration',
'Coding Standard',
'User stories',
'Pair programming'.

None of these terms have anything to do with the academic discipline of Software Engineering. SCRUM breaks down the production process into short phases known as 'Sprints'. Each 'Sprint' in theory only lasts two weeks. But in practice, some Software Developers make this last up to 4 weeks or more. It begins with a set of targets, sometimes called 'deliverables'. And attached to each 'deliverable' is a total time estimated to complete it. This time is reduced each day as progress is made. And the total time left for all 'deliverables' is plotted on a chart known as a 'Regression Chart'. The chart plots time spent on the 'Sprint' on the X-axis, against time left for all 'deliverables' on the Y-axis. And in theory, it should show a straight line, slopping down, from some arbitrary level, to 0, from left to right, as the end of the 'Sprint' approaches.

But in practice, the line is never straight, nor does it reach 0. The estimates to complete the 'deliverables' are nothing more than educated guesses. And this relies on an iterative process to correct. SCRUM also allows the phases of the classic software production cycle to be overlapped, as different groups work on different 'deliverables' at the same time. SCRUM has no provision for even prioritising 'deliverables' and it places no value on documentation.

Therefore, SCRUM too bears the characteristics of a Software Evolution Process. It is both an iterative process and non-linear and places no value on written designs.

Less and less effort is spent on fixing original design flaws; more and more is spent on fixing flaws introduced by earlier fixes...As time passes, the system becomes less and less well-ordered. Sooner or later the fixing ceases to gain any ground. Each forward step is matched by a backward one. Although in principle usable forever, the system has worn out as a base for progress.

Source: The Mythical Man Month: Essays on Software Engineering, 20th Anniversary Edition © 1995, Addison-Wesley. Frederick P. Brooks

2 The Laws

2.1 I - Continuing Change

An E-type program that is used must be continually adapted else it becomes progressively less satisfactory....

2.2 II - Increasing Complexity:

As a program is evolved its complexity increases unless work is done to maintain or reduce it...

2.3 III - Self Regulation

The program evolution process is self-regulating with close to normal distribution of measures of product and process attributes...

2.4 IV - Conservation of Organisational Stability (invariant work rate)

The average effective global activity rate on an evolving system is invariant over the product life time...

2.5 V - Conservation of Familiarity

During the active life of an evolving program, the content of successive releases is statistically invariant...

2.6 VI - Continuing Growth

Functional content of a program must be continually increased to maintain user satisfaction over its lifetime...

2.7 VII - Declining Quality

E-type programs will be perceived as of declining quality unless rigorously maintained and adapted to a changing operational environment...

2.8 VIII - Feedback System

E-type Programming Processes constitute Multi-loop, Multi-level Feedback systems and must be treated as such to be successfully modified or improved...

Source: Laws of Software Evolution Revisited © 1997, M. M. Lehman

The stated, accepted philosophy for systems development is that the development process is a well understood approach that can be planned, estimated, and successfully completed. This has proven incorrect in practice. SCRUM assumes that the systems development process is an unpredictable, complicated process that can only be roughly described as an overall progression. SCRUM defines the systems development process as a loose set of activities that combines known, workable tools and techniques with the best that a development team can devise to build systems. Since these activities are loose, controls to manage the process and inherent risk are used. SCRUM is an enhancement of the commonly used iterative/incremental object-oriented development cycle.

Source: The Scrum Development Process © 2024. Scrum.org

Extreme Programming (XP) is a software engineering methodology, the most prominent of several agile software development methodologies. Like other agile methodologies, Extreme Programming differs from traditional methodologies primarily in placing a higher value on adaptability than on predictability.…Extreme Programming was created by Kent Beck, Ward Cunningham, and Ron Jeffries during their work on the Chrysler Comprehensive Compensation System (C3) payroll project. Kent Beck became the C3 project leader in March 1996 and began to refine the development methodology used on the project. Kent Beck wrote a book on the methodology, and in October 1999, Extreme Programming Explained was published. Chrysler cancelled the essentially unsuccessful C3 project in February 2000, but the methodology had caught on in the software engineering field.…

Source: Extreme Programming © 2007. Wikipedia. The Free Encyclopedia

[A client came in with the idea of the game for them to create for the military and large companies. And made it clear that they would be using a SCRUM PRODUCTION PROCESS, with SPRINTS of 2 weeks to develop it.]
*   Game Producer #1: "Yeah! That is the only danger [in the project] that I can see out of the whole thing"*
*   [making impromptu speculation about the feasibility of the client's project.]*
*   Client #1: "…raise information security awareness …increase the understanding of the security team. …we would want a high level design, then a 2 week Sprint … within the first 2 weeks we have agreed a high-level design. This is not a deliverable you will have to give us…"*

Source: A typical Diary of a Software Evolution Process of Slippery Games Inc.
Anonymous. February 2014

6.18 PRACTICAL APPLICATION (OF BIOLOGICAL EVOLUTION)

Advocates of the application of the theory of Biological Evolution, to software production, can be found in many quarters of the Software industry. Even those who develop software, which helps other industries construct intricate designs, and analyse these before production, believe that a less analytical, more evolutionary approach, suits the Software industry. These include John Walker who, together with other staff at Autodesk, developed one of the most popular Computer-Aided Design tools, AutoCAD. This tool has been used by other industries, to comprehensively design and analyse products prior to production. A derivative of this tool, 3DS Studio MAX, also became the most popular tool used by Game Artists in the Computer Games industry.

Nevertheless, by adopting the theory of Biological Evolution, its advocates neglect or deny any ethical dimension to a software production process. They deny any ethical responsibility, on the part of the Software Developer, to produce their best effort. They deny any responsibility for the production process to begin and end with the needs of the customers.

Instead, the theory of Biological Evolution requires a Software Developer to only produce makeshift products. Each attempt has to get out to the market as quickly as possible, to find out which way the market is heading. It does not have to meet the customers as far along that path as possible; merely find out where that path is leading. And all the Software Developer has to do is keep promising that the next version will go further down that path.

But, of course, the next version will be another makeshift product. If the theory of Biological Evolution suggests that a Developer may forego the analysis of the very first version of the product, and just let the natural selection take its course, why would the Developer perform any analysis for any subsequent versions? So the customers will continue to wait, while the Software Developer continues to get money for their makeshift efforts.

In this unethical production process, the needs of the customers are secondary to the survival of the Software Developer. The process does not begin with the needs of the customers. Nor does it end with them. Instead, it begins with the need for the Software Developer to quickly secure some emerging consumer market. And, theoretically, it never ends as the product can keep mutating, assimilating more and more features.

In practice, however, the product quickly becomes too unwieldy for the Software Developer. And other competitors, with relatively newer products, react faster to the market than the Developer. So the only way open for the Developer to survive is by assimilating these competitors, their products or features of their products. This, in turn, raises another set of ethical issues, which those who advocate the application of the theory of Biological Evolution do not even recognise. Should large Software Developers with unwieldy legacy products be allowed to consume smaller Software Developers with new innovative products? What happens to competitiveness and innovation in the market if you allow this?

At this point, it becomes self-evident which species survival depends on the application of the theory, to software production. The species is not some set of products, of a Software Developer, which share some common ancestry. Nor is it some set of customers, of the Developer, who share some common need. It is the Software Developers themselves.

I use the word "evolution" a lot because I believe it's central to understanding how markets really work, how technologies emerge and mature, and how actual products are developed in the real world. In the early days of Autodesk, I didn't even try to guess which product would succeed–I knew I wouldn't have a hope of making such a prediction accurately. But I was pretty confident we could bat .200–that at least one out of five products we chose would succeed in the market. Then, and only then, would we focus our efforts upon the winner.

Think of it as evolution in action. We, as product developers, are creating new species, almost as blindly as the shuffling of genes, with the market—our customers–performing the winnowing process of selection. As in biology, there's no way to know how well something will work without trying it. Yet once it gets out there, you learn pretty quickly whether it was a good idea or just plain dumb....

That's why it's ever so important to get a product into the field early and to have a rapid and responsive development and upgrade program. The first product in a category benefits from the feedback of customers and can quickly begin to converge toward meeting their requirements, often growing in directions not remotely anticipated in the original design.

...it explains why large, mature products tend to be messy and complicated, because they have accreted, over the years, a large number of features, each requested by and valuable to, a set of customers.... Only when a customer ceases to believe that the product he already owns will meet his needs in the future does he goes shopping for a replacement.

All this seems so obvious to me that I rarely bother trying to explain it, and yet the process by which products are proposed and developed in many organisations, including Autodesk, seems diametrically opposed to this evolutionary philosophy. Instead, we do market research (asking people what they think about something that does not exist) in order to make a detailed design, forecast market acceptance in advance, then build the product all-up to be perfect from the start.

Source: The Autodesk File, Bits of History, Words of Experience
© 1994, John Walker

6.19   OPEN-ENDED PROCESS (OF BIOLOGICAL EVOLUTION)

The extinction of a species is not the end-point or goal of Biological Evolution. The goal is survival; survival of the fittest.

According to the theory of Biological Evolution, when a species evolves into many subspecies, it is in order to survive. If one of the subspecies becomes extinct, that does not mean that the process of Biological Evolution has ended. The other subspecies, and the parent species, survive and continue to evolve. Even within the limited perspective of the subspecies which became extinct, there would have been no way of predicting when that species would reach that point. Nor would there have been a requirement for the species to reach that point. Thus, the process would still be open-ended.

6.20   ART PIPELINE

A manual, distinct sub-process of the production of Computer Games, for generating artwork. The Pipeline is created by and for Game Artists. And it usually begins with a separate document, sometimes known as a media design, if not the game design. This document describes the sub-process, its objectives, the techniques, tools and other resources that would be used to achieve these objectives.

'Texture Artist
Founded in 1993 this Ontario studio proudly ranks as one of the world's top development studios in the interactive entertainment industry...
Candidates Will

- *Create photo-realistic textures from scratch as well as photo source materials.*
- *Use creative and efficient UV mapping to apply textures to 3D geometry.*

- *Create normal maps from high-poly geometry.*
- *Work within, and optimize for, real-time memory and shader constraints.*
- *Use proprietary tools and work within an established asset development pipeline.*

Candidates Must Have:

- *Extensive environmental texturing experience.*
- *Experience with next-gen art pipelines and production preferred.'*

Source: Game artist and animator jobs © 2007. Datascope Recruitment

The art pipeline is a term used to describe the entire process of creating and implementing art for a particular project, most commonly associated with the creative process for developing video games. In an era of high profile video games, wherein the creative energy of the teams and the budgets for projects surpass even some Hollywood blockbusters, graphics are ever-improving and an increasingly important selling point. Video Game developers employ extensive teams of artists to carry a project's artistic goals through from the conceptual stage to the final release. A fully realized game asset, whether it is a character, background, building, object, or animation, is created in a deliberate process with different artists working on and contributing separate aspects in a step by step process to the final product.

Source: Art Pipeline © 2007. Wikipedia. The Free Encyclopedia

6.21   BUILD PIPELINE

An automated, distinct sub-process of the production of Computer Games, for periodically building and testing the game to ensure no errors have been introduced into the building process, due to rapid changes.

There may be between 80 and 100 changes going into the game a day, 3 and 4 changes an hour or 1 change every 15 minutes, for a project with about 60 staff.

The system reports any errors through e-mails to potential authors who it suspects made changes that produced the errors. But the system does not really know who or what produced the errors. Since the system deals in aggregates. The system reports the aggregate number of errors that were produced, the aggregate number of Software Users who submitted changes and the aggregate number of changes.

For example, the system will tell you that at 12.00 a.m. today, there were 12 errors reported, after 20 Users submitted changes and there were 40 changes. Now the number of files involved in these 40 changes could be anywhere from 40 files to 100 files. And the number of lines in these files could be anywhere from 40 lines to 10,000 lines.

Since the system cannot identify the source of the errors, it just sends an e-mail to all those it suspects caused the errors. The suspects are usually the last set of Software Users who submitted files to a Software Repository which the system used to build the latest version of the game. The automated system which does this is sometimes called Continuous Integration or CI.

The system may also automate the deployment or release of the game to the Software User i.e. players, including

1. announcing the latest version of the game through some Web Server
2. testing the game before release
3. creating custom versions of the game for different Operation Systems or computer hardware or 'platforms' before release to those 'platforms'
4. creating special software to control the uninstalling of the old version of the game and installing the latest version.

The systems which automate the deployment or release of the game are called Continuous Deployment or CD.

Many Software Developers confuse CI and Continuous Deployment (CD) systems. But most Software Developers only have a CI system. Since they typically work for third parties who control when and how the game is announced. Some Software Developers who work for themselves or have completed the production of a game for a third party do have a Continuous Deployment (CD) system.

The Game Programmer responsible for these CI or CD systems is called a Build Engineer. The Build Engineer often gets scapegoated for the errors the CI or CD systems reveal. Due to the rapid changes going into the game, from several sub-processes running in parallel in the Software Evolution Process, no one can keep up with the changes. Apart from the CI or CD systems.

These systems are often the first to put all of these changes together and also the first to reveal the errors and Bugs (euphemistically sometimes called 'build issues') which arise when you do so. The Build Engineer is in theory only responsible for the 'build stability' and chasing up the Software Users who made changes that caused the errors. But in practice this requires the Build Engineer to investigate these errors to find out who is responsible. And that investigation is hampered by the inscrutability of the Software Evolution Process. Nevertheless the failure to scrutinise the process and correctly identify who is responsible for errors ends up being attributed to a flaw in the CI or CD system or the Build Engineer. When it is in fact a flaw in the Software Evolution Process.

BUILD ENGINEER

... looking for a talented and passionate Build Engineer to join our team... In this role, you will be responsible for the development, maintenance and optimisation of all our build pipelines....

RESPONSIBILITIES

- *Developing, maintaining, and optimising our multi-platform build pipelines.*
- *Managing and communicating the flow of changes in source control across multiple environments.*
- *Working with developers to resolve and validate merge conflicts across multiple environments.*
- *Communicating and coordinating releases (lockdown approvals, patches, etc.) with colleagues of various disciplines.*

- *Identify opportunities to improve and automate our processes.*
- ....
- *Working with QA to help triage build and runtime issues and their potential solutions.*
- ....

EXPERIENCE GUIDELINES

- *Passionate about playing and building video games!*
- *Practical experience of version control systems (ideally Perforce and Git).*
- *Practical experience in constructing and managing build pipelines for multiple environments and platforms.*
- ...
- *Proficiency writing and maintaining build scripts.*
- *Familiarity with CI/CD concepts and tools such as TeamCity.*
- ....
- *Attention to detail and interest in catching build issues.*
- *A proactive attitude toward improving the build/release processes...'*

Source: A typical Build Engineer Job Advert from Slippery Games Inc. Anonymous. 2024

6.22   EVOLVE SOFTWARE WITHOUT DEGENERATING

M. M. Lehman, who gave the Software Evolution Process its name, has always found the process degenerative and denied any link between it and generative biological evolution.

The reason he started examining the process in the first place was precisely to discover what factors determined how it degenerated. However, to date, he has not formulated a final theory to explain the phenomenon. Although many are quick to link his theory with the theory of Biological Evolution, by Charles Darwin, he is not. This is the fundamental difference in how Lehman uses the phrase Software Evolution Process, and its popular interpretation. Lehman uses it as a basis for formulating a theory. As part of this formulation, he wrote 8 laws to explain Software Evolution.

Others have subsequently conflated Software Evolution with the theory of Biological Evolution. They assumed to have an understanding of the laws of Software Evolution, from their understanding of the theory of Biological Evolution. But they have not examined these laws carefully. Thus, the laws of Software Evolution, at least their understanding of these laws, have assumed the same level of credibility as the theory of Biological Evolution. They have assumed that these laws had some basis in nature. And as such, they find it incredulous to accept a criticism of Software Evolution, as distinct from a criticism of the theory of Biological Evolution. They have assumed that these laws could not be changed, like the laws of nature. And as such, a production process which did not follow the laws of Software Evolution, from the beginning, would fail.

But, firstly, Lehman himself is changing these laws and has not finalised his theory. Secondly, the origin of Software Evolution was a study of software in its operational and maintenance phase, of the software production life cycle. It did not come from an examination of the entire software production life cycle, from the beginning.

> *For Lehman, the place to look is within the software development process itself, a system Lehman views as feedback-driven and biased toward increasing complexity. Figure out how to control the various feedback loops – i.e. market demand, internal debugging and individual developer whim – and you can stave off crippling over-complexity for longer periods of time. What's more, you might even get a sense of the underlying dynamics driving the system....*
>
> *...At a time when lay authors and fellow researchers feel comfortable invoking the name of Charles Darwin when discussing software technology, Lehman holds back. "The gap between biological evolution and artificial systems evolution is just too enormous to expect to link the two," he says...'*
>
> Source: A unified theory of software evolution © 2003, Salon.com. Sam Williams

6.23   FEEDBACK FROM THE SOFTWARE USERS

Virtually all commercial software licences exclude the software users from the production process.

In theory, although the laws of Software Evolution require feedback from the software users, the users have not been involved in the Software Evolution Process, as practised commercially. The maintenance of commercial software (i.e. software not used in-house by a company) does not officially occur. Virtually all commercial software licences explicitly deny any warranty and do not claim any suitability of the software for any purpose. Any maintenance is done at the user's personal cost and risk.

These software licences, including those for Computer Games, exclude the end-users from the production process. This runs contrary to the laws of Software Evolution that require, in theory, the production process to be based on feedback mechanisms. In practice, none of the mechanisms in the Software Evolution Process have anything directly to do with the end users. Except, that is, the influence of those professionals involved in the process, who claim to be the users' proxy.

Even if the users were involved in the process, their ability to provide feedback would be muted. In theory, each iteration of the Software Evolution Process should be progressive, based on information fed back by the software users. That is each step should be making the software more robust and bringing it closer towards the final product. In practice, the negligence with respect to any designs or plans, which marks the beginning of the Software Evolution Process, carries on throughout the rest of it. So the components of the software which were meant to produce information, by for example displaying errors or messages which the software users could understand, would also be neglected. And no useful information would be fed back into the process by the users.

Hence, each step of the process, in practice, would be blind. It could just as likely be regressive, making the software less robust and moving it further away from the final product, as being progressive. Indeed, this alternation between regression and progression would be reflected by the names given to the various versions of the software that appear during the process. The regressive versions would sometimes be referred to as unstable or development versions. Whereas the progressive ones would be known as stable versions. In the end, the regressions occur more often in practice, than the progressions, inevitably bringing the Software Evolution Process to a halt.

The software industry is not very well known for its warrantees, but is much more famous for its legal disclaimers absolving software firms for any and all liability for its products. One such unfortunate and sweeping disclaimer follows:

Cosmotronic Software Unlimited Inc. does not warrant the functions contained in the program will meet your requirements or that the operation of the program will be uninterrupted or error-free.

However, Cosmotronic Software Unlimited Inc. warrants the diskette(s) on which the program is furnished to be of black color and square shape under normal use for a period of ninety (90) days from the date of purchase

Note: In no event will Cosmotronic Software Unlimited Inc. or its distributors and their dealers be liable to you for any damages, including any lost profit, lost savings, lost patience or other incidental or consequential damage.

We don't claim Interactive EasyFlow is good for anything - if you think it is, great, but it's up to you to decide. If Interactive EasyFlow doesn't work: tough. If you lose a million because Interactive EasyFlow messes up, it's you that's out of the million, not us. If you don't like this disclaimer: tough. We reserve the right to do the absolute minimum provided by law, up to and including nothing.

This is basically the same disclaimer that comes with all software packages, but ours is in plain English and theirs is in legalese.

We didn't really want to include a disclaimer at all, but our lawyers insisted. We tried to ignore them, but they threatened us with the shark attack at which point we relented.45

Another extraordinary aspect of software marketing is the fact that the user generally pays for software updates. In other words, even if the product is faulty or needs amendment, the user pays the software supplier to provide more correct versions.

[Forester]

Source: Ethics in Military and Civilian Software Development
© 1999, Sam Nitzberg

6.24   GAME PRODUCER

An employee of a games company in charge of the overall production of a game, from getting its financing, through its analysis, design, implementation, and testing, to its release.

6.25   GAME DESIGNER

An employee of a games company responsible for game designs.

6.26   GAME TESTER

An employee of a games company who tests the software at the end of its production.

6.27 GAME PROGRAMMER

An employee of a games company who writes the software for games.

6.28 GAME ARTIST

An employee of a games company who makes 3D models, 2D images and animations used in a game.

6.29 SOUND DESIGNER

An employee of a games company who creates and records the sound and music played in a game.

6.30 POST MORTEM MEETING

A meeting conducted at the end of a software production process, by the staff involved, to retrospectively examine the pros and cons. And to decide the lessons to be learnt from the experience. The meeting comes from the academic study of Project Management, not software engineering.

What Is a Post-Mortem Meeting?

A post-mortem meeting is a formal discussion that occurs at the end of a project. In the meeting, the project team discusses what went right and wrong and uses that information to make process improvements for future projects

> Source: How to run a Post-mortem meeting (c) 2024. Smartsheet Inc.

Please add any comments related to the working with other teams here:

"You are awesome!!"

Publishing reshuffling the priorities of bugs to suit their needs is unhelpful i.e. "They make their personal issues top priority so you end up fixing a wrong 'cuddle a teddy bear' animation, rather than game breakers."

Design should be more assertive and not allow Publishing and execs to change designs.

"[Last minute changes were reasonable...] if they weren't camouflaged as a feature request."

"xbox save system. This was a last minute bug request, which actually turned out to be a whole rework of it."

"...kudos to our compliance teams ([Redacted] and others). They have been very helpful in answering questions."

R—— Tech sending bugs straight back, requesting that we verify the bugs are actually for them, when that should really be their job.

"Everyone working on [Redacted] is generally quite nice so no complaints here!"

"Being part of [Redacted]Systems I sometimes find myself not being informed of being part of discussions about features directly involving our systems."

"Sometimes g—tech don't like us sending things their way and we have to fix their bugs."

"Members from other teams who chipped in in the run up to [Redacted] were spot on."

"Production have been spot on in almost all interactions."

"Additional design or dev bugs coming in from publishing are sometimes really poorly scoped or have insufficient information."

"More fleshed out, more timely designs for the tasks would be appreciated."

"Last minute change requests have been a little extreme in a few cases."

"G—- tech were good!"

....

1. ***Were you happy with the kinds of tasks that were assigned to you?***

    "Yes!"

    "Can't complain"

    "Yes. My work was pretty much always scoped to my expertise."

    "Yes, most of the time, but sometimes bugs like to hide or be difficult to track."

    "Most of the time I was bug fixing monkey, but retrospectively it was ok... but frustrating"

    "Yes, happy to take on more"

    "I'm happy with any task, I like challenges."

    "Yeah for the most part."

> *Source: A typical Post Mortem Document of Slippery Games Inc. 2017. Redacted. Anonymous.*

1. *Overall, how do you feel [Redacted] went?*

    It could have been managed better. It was initially going to be released around September? But then it was moved back to December? And then it was moved again back to March? And then it was moved back again to June? Missing so many deadlines seems bad for the reputation of the company

2. *Which aspects went well during the development of [Redacted]?*

    The optimisation which some how managed to improve the performance of [Redacted] to the point where the Producers ([Redacted[?) were satisfied with it. The User Interface seemed to go through a lot of changes and iterations but somehow the people working on it managed to keep up with the changes.

3. *What expectations did you have about working on [Redacted], that did or didn't happen?*

    I was expecting [Redacted] to be released in December. But that did not happen. I expected to rely on other people's expertise to help me understand the project. And they did help when they could especially ...

4. *What frustrations did you encounter working on [Redacted]?*

    There was little sense of a Game Design, and how far the production process was to completing that Game Design. It seemed that everything was in a state of flux, features were being added or removed, without any announcement.

5. *What one thing would be most important to improve for our future projects? This can be something you have already mentioned that you would like to call out as the most important thing to you*

    There should have been more show-and-tell, to give people an idea of what other people were working on and how far they had progressed.

6. *Anything Else?*

    There was little or no documentation of work done. A lot of communication seemed to rely on informal verbal communications, when other people may be busy and do not have the time to stop their work and explain. And this is also a flaw in the [Redacted] Engine. This Post-Mortem is far too late to feedback into other Projects which are about to finish or half-way through.'

Source: A typical Post Mortem Document of Slippery Games Inc. 2020. Redacted. Anonymous

6.31   TOWER OF BABEL

In his book, 'The Mythical Man Month', Frederick P. Brooks compared the confusing language that arises in the Software Evolution Process to the building of the Tower of Babel.

In Chapter 7, entitled 'Why Did The Tower of Babel Fail?', he recognised how the quality of language deteriorates in the process and along with that the communication within the project. As different factions adopted slightly different languages to describe the same process, everyone found it harder and harder to communicate: just like in the Bible when God stopped the building of the Tower of Babel, by simply making each of the builders speak different languages.

According to the Genesis account, the tower of Babel was man's second major engineering undertaking, after Noah's ark. Babel was the first engineering fiasco...

Well, if they had all of these things, why did the project fail? Where did they lack? In two respects—communication, and its consequent, organization. They were unable to talk with each other, hence they could not coordinate. When coordination failed, work ground to a halt. Reading between the lines we gather that lack of communication led to disputes, bad feelings, and group jealousies. Shortly the clans began to move apart, preferring isolation to wrangling.

Source: The Mythical Man Month: Essays on Software Engineering, 20th Anniversary Edition © 1995, Addison-Wesley. Frederick P. Brooks

6.32   PRIVATE CONVERSATIONS

Some isolated observers may view the private conversations and the decisions which come out of these, as beneficial to the productivity of a Software Evolution Process. But that would be a mistake. They may see these conversations as a sign that, at least, some of the staff have superior knowledge of how the software works. And they can use this to make incisive decisions that advance the project forward; a project which would otherwise stagnate if the conversations involved all of the staff.

But such observations would fail to consider other alternative explanations for this apparent superior knowledge. Namely, the exclusivity of these conversations stemmed from a serious breakdown in communication, which had occurred early on in the project. And that the staff who indulged in these conversations did not have superior knowledge. They were merely using a language which the rest of the staff could not understand.

Even worse, such observations would fail to recognise that, in a collaborative team effort, there could be no superiority. A team could only function if all of its members were partners. And if they were partners, they would all have to be equals.

If a subset of the staff controlled the destiny of a Software Evolution Process then, to the members of that small team, the rest would be redundant. As such they could not, in all sincerity, think the rest were equals. And if the rest were not equals, they would have no reason to collaborate with them. The notion of more effective small teams within larger ones is as much an oxymoron as the notion of first amongst equals.

An excellent example of how the Software Evolution Process makes a small subset of a team involved in the Process seem to have superior knowledge to the rest of the team, but in fact they do not have superior knowledge and merely use a different natural language, can be seen when you examine modern Game Engines. Take, for example, the Unity Engine and Editor and the Unreal Engine and Editor.

Both these tools were made to address the same problem i.e. building computer games on almost exactly the same set of modern computer hardware. And yet the language used by both sets of tools to describe this same problem, and the same process for solving this problem is very different. Now it may seem that a small subset of a team which is familiar with the Unity Engine has superior knowledge to the rest who are familiar with the Unreal Engine when the company they work for requires them to use the former. Likewise, it may seem that a small subset of the team which is familiar with the Unreal Engine has superior knowledge to the rest who are familiar with the Unity Engine or other Game Engines when the company they work for require them to use the Unreal Engine.

But that perception is false. The subset does not have superior knowledge even though they seem to be more productive. The subset is simply using a language and terms that the rest of the team does not readily understand. You can see the difference between the two languages for the two Game Engines in Table 6.1.

TABLE 6.1
Unreal Engine for Unity Developers

Category	Unity	Unreal Engine
Gameplay Types	Component	Component
	Game Object	Actor
	Prefab	Blueprint Class
Editor UI	Hierarchy Panel	World Outliner
	Inspector	Details Panel
	Project Browser	Content Browser
	Scene View	Level Viewport
Meshes	Mesh	Static Mesh
	Skinned Mesh	Skeletal Mesh
Materials	Shader	Material, Material Editor
	Material	Material Instance
Effects	Particle Effect	Effect, Particle, Niagara
Game UI	UI	UMG (Unreal Motion Graphics) and Slate
Animation	Animation	Skeletal Mesh Animation System
	Mecanim	Animation Blueprint
	Sequences	Sequencer

(Continued)

TABLE 6.1 (*Continued*)
Unreal Engine for Unity Developers. Comparisons of the terminology of the unreal engine and the terminology of the unity engine.

Category	Unity	Unreal Engine
2D	Sprite Editor	Paper2D
Programming	C#	C++
	Script, Bolt	Blueprint
Physics	Raycast	Line Trace, Shape Trace
	Rigidbody	Collision, Physics
Runtime Platforms	iOS Player, Web Player	Platforms

Comparisons of the terminology of the unreal engine and the terminology of the unity engine.
Source: Unreal Engine for Unity Developers. 2024. Anonymous.

And it takes the rest a long time to pick up each language. In part because there is little documentation to explain both Game Engines, which is a consequence of the Software Evolution Process used to make these Engines. And in part because those who are familiar with one Engine have not got the time, inclination or eloquence to explain the terms to those who are not. And in part because the Software Evolution Processes, using these Engines, are themselves inventing new words, redefining words and adding to the overall confusing language of the Process.

A small subset of the design team with superior application domain knowledge often exerted a large impact on the design...the small, but influential, design coalitions that developed on numerous projects represents the formation of a small team in which collaboration was more effective. This decomposition of a large design team into at least one smaller coalition occurred when a few designers perceived their tighter, less interrupted collaboration would expedite the creation of a workable design.

Source: A Field Study of the Software Design Process for Large Systems © 1988, Association of Computing Machinery Inc.
Bill Curtis, Herb Krasner and Niel Iscoe

6.33   RESEARCH STUDIES

Studies conducted for the International Business Machines Corporation (IBM) and the Microelectronics and Computer Technology Corporation (MCC) showed that improved communication had a beneficial effect on productivity.

Study participants indicated that the improved communication among community members contributed to successfully executed projects, increased new business, and product innovation.

Source: Understanding the Benefits and Cost of Communities of Practice © 2002, Association of Computing Machinery Inc. David R. Millen, Michael A. Fontaine and Michael J. Muller

Many techniques were used to organize and communicate a shared system model. Successful projects usually established common representational conventions to facilitate communication and to provide a common reference for discussing system issues. From a team perspective, this sort of representation was valuable as a common dialect for project argumentation, rather than as a basis for static documentation.

System engineer: The ER diagram means that everybody speaks the same language. Developers, designers, human performance people, we all use the same language...It was 6 months or so before it settled down, but once it did, we could resolve all problems in terms of the diagram.

Source: A Field Study of the Software Design Process for Large Systems © 1988, Association of Computing Machinery Inc. Bill Curtis, Herb Krasner and Niel Iscoe

6.34   LACK OF A PLAN (IN SOFTWARE EVOLUTION)

Instead of the plan (i.e. the game design and technical design) providing for contingencies, the plan itself becomes a contingency, i.e. a future event which cannot be predicted with certainty.

Since the plan becomes a future event, in the Software Evolution Process, the process begins with just a vague set of points, sometimes called a 'Shopping List' or a 'Wish List', listing features that may be in the final plan or product (i.e. the game).

And a lot of uncertainty remains throughout the process, right till the very end, about what the final plan will be.

J—- [Redacted] wanted some kind of tool that could be used to write a quick game, based on HTML5 in one day. This was in response to S—- [Redacted] receiving an E-MAIL from the BBC asking him to confirm what kind of technology F—- P—- [Redacted] be using, including HTML5.

One suggestion was to use FLAMBE game engine. ... C—- [Redacted] thought it may take a few Fridays to do.

Source: A typical Diary of a Software Evolution Process of Slippery Games Inc. 2014. Redacted. Anonymous

J—- [Redacted] shared the latest version of his SOFTWARE DESIGN for u—- 2.0 [Redacted] on GOOGLE DRIVE:...

It was very vague. It had no introduction. But instead went straight into giving a vague technical description of the C++ CLASSES that will be used, the tools, and the SOFTWARE PRODUCTION PROCESS for building it. It was all arranged in a random manner.

The document was edited to include an introduction...

He was keen to get started making u—- 2.0 [Redacted] despite the vagueness of the document. He wanted to set up the SOFTWARE REPOSITORY for it straight away, although he had no clear idea of what was going to go in that REPOSITORY apart from the files of A—- SOFTWARE LIBRARY [Redacted] for developing software based on F—- P—- [Redacted]...

J—- [Redacted] wanted the REPOSITORY set up. So that it would include files (CMakeLists.txt) which you could use with the CMAKE command to build the u—- 2.0 [Redacted] on different COMPUTER HARDWARE and OPERATING SYSTEMS straight away. Even though u—- 2.0 [Redacted] had no written design, he had no idea what he was going to implement in it, apart from one or two "core" MODULES in his head, he had no clear vision about where u—- 2.0 [Redacted] was going.

Initially, before today, he said that it would only implement a subset of features of F—- P—- [Redacted]. But today he said it would implement the full set of features. After he read the amendements to his crude the design, which had been amended to include an introduction which explicitly spelt out the initial goals of u—- 2.0 [Redacted], he changed his mind. He could see that u—- 2.0 [Redacted] was not going to be all that he had promised to his customers. So he forced himself to commit to implementing all of the features of F—- P—- [Redacted].

In the end, the REPOSITORY was not set up for the CMAKE command today. He claimed to be disappointed by the progress made that day. But seemed delighted when the offer was made to continue work on it tomorrow. He said then "we could crack on on Monday" or words to that effect. He seems eager to begin something he has no desire to spell out in a design, and therefore has no clear vision where he is going to. And he gets upset when people expose this ignorance by insisting for more details in some kind of written design. He kept repeating this when asked for clarification as to what the CMAKE command was meant to build:

[–/–/2014 –:–:–] J—- [Redacted]: heres the cmake setup required:
u—-[Redacted].lib (static lib)

- *compile src/u—-[Redacted]*

u—-[Redacted]test.exe (test executable)

- *compile test/u—-[Redacted]/*
- *compile thirdparty/UnitTest++*
- *compile   projects/xcode/u—-[Redacted]2-proto-dev/u–2-proto-dev/main. cpp*

As if that was meant to mean something outside of his own head. As if that explained the goals and vision for the project.

Source: A typical Diary of a Software Evolution Process of Slippery Games Inc. 2014. Redacted. Anonymous

6.35   DEFAULT PROCESSES (BASED ON NEO-DARWINISM)

Other industries tend not to make explicit references to the theory of Biological Evolution, through the names of their default processes. They are more pragmatic about what they call these processes. And, but for the efforts of M. M. Lehman, the Software industry would not have adopted the Software Evolution Process, to refer to their default processes either.

Instead, they would have been content, as has been the case with other industries, to use 'trial and error' or 'prototyping'. Nevertheless, all of these default processes, both in Software and other industries, rest on the same two pillars as Neo-Darwinism. Namely, theoretically, each one rests on the small, incremental, evolutionary, growth of a product, accompanied by some form of natural selection.

In other industries, this growth and natural selection would informally be directed by the experience and intuition of those involved in the process. So you would rarely come across any explicit references to Neo-Darwinism. But some direct this growth and natural selection through a more formal method. And in the description of these methods, you would find references to Neo-Darwinism. These include, for example, The Theory of Inventive Problem Solving (or TRIZ in its Russian abbreviation), which has been applied in Manufacturing industries, Biomedical Research and Medicine, amongst others.

> *Trial and error still plays a key role in product development. The answer is not to avoid mistakes but to make them early and often.*
> *...*
> *Designers dislike the phrase "trial and error" - it sounds so uncontrolled and wasteful...But some trial and error remains at the heart of new product development. That's especially true for smaller companies who are not trying to protect an established product niche but hoping to create something entirely new. In doing so, design engineering shares a methodology with people engaged in the arts.*
> *The novelist writing a book, the sculptor chiseling a statue, the musician composing a concerto, the engineer designing a bridge, and the theater director giving her actors the first crude stage blocking all allocate their resources so that improvement in their product is by successive approximations," wrote Billy Vaughn Koen in his book "Discussion of The Method: Conducting The Engineer's Approach To Problem Solving."*
> *Koen sees the engineering method everywhere: All creative endeavors are a form of problem solving in which perfection is not possible...Picasso's early sketches for his famous painting "Guernica" is "absolutely identical to what the engineer is doing - getting to the goal by successive approximation..."*
> *Lenny Lipton is the CEO and principle researcher of StereoGraphics, a pioneer in 3D video displays...*
> *"Day to day, most of what I do is still trial and error," he says. "Obviously, you need some wisdom, based on experience. But mostly, I try different things, find out what works, and stop doing what doesn't work. Trial and error is a pragmatic form of hope."*
>
> *Source: Thinking in Prototypes © 2005, Design Continuum. Bart Eisenberg*

Effective and efficient development of new generations of products and processes is the mighty weapon in the competitive struggle. Presently, there is no structured methodology to perform this extremely important activity and the prevailing approach is the "trial and error,"...

The theoretical foundation of the TRIZ technology forecasting is a set of the Laws or Prevailing Trends of Technological Systems Evolution revealed by analysis of hundreds of thousands of invention descriptions available in the world patent databases...

The Laws of Evolution reflect significant, stable, and repeatable interactions between elements of technological systems and between the systems and their environment in the process of evolution...

Thus, the main benefits of the TRIZ forecasting are the following:

- *TRIZ forecast means developing conceptual designs of new systems. In other words, TRIZ forecast shows not only what will happen, but also how to achieve the desirable results.*
- *Higher accuracy of the forecast, since it is based on the Laws of Technological Systems Evolution.*

- *Detection of the point in time when development of the present technology should be abandoned and new directions should be explored.'*

6.36   NEO-DARWINISM

The synthesis of a modern theory with the theory of Biological Evolution by Charles Darwin. Usually, this refers to the synthesis of genetics with Darwin's theory.

When he conceived his theory, Darwin was not aware of the role that genes played in determining the characteristics of animals and plants. Genes are small parts of molecules, found in the cells of living animals and plants. The components of each gene determine the appearance of the animal or plant, its eyes, hair, hands, nose, legs, colour etc.

Although Darwin proposed, in his theory, that animals inherited traits from parents, he did not know that these came from the genes of the animals. He also did not believe that random mutations were responsible for the diversity within animals. Instead, he believed that the habits of the parents progressively, physically changed the bodies of these animals. And these habits and physical traits were, in turn, inherited by the children. And this was responsible for the diversity.

It was only later, after Charles Darwin had published his theory in 1859, that it was discovered, in 1865 by Gregor Mendel, that the physical changes of animals or plants could not be inherited by the children of animals or plants. Even later still, in 1910, 28 years after Darwin died, Thomas Hunt discovered that genes were the device which controlled, at least, physical inheritance. And shortly after that, the idea of random mutations was added to the modern concept of Biological Evolution. Random errors that occurred, for example, during the copying of genes or due to radiation, within cells, were put forward as being responsible for these random mutations. And these mutations were responsible for the diversity of animals.

But, to date, all attempts to artificially replicate such mutations have been unsuccessful. And the results have either been benign, only helping protect against some diseases, but without changing physical appearance. Or the results have been negligible or degenerative.

Nevertheless the attempts to synthesise modern theories and Darwin's original theory, to demonstrate the feasibility of Darwin's theory have produced Neo-Darwinism. And Software Evolution is just another instance of Neo-Darwinism.

neo-Darwinism
noun
neo-Dar·win·ism |\-'där-wə-ˌniz-əm
\
Medical Definition of neo-Darwinism
: *a theory of evolution that is a synthesis of Darwin's theory in terms of natural selection and modern population genetics'*

The importance he placed on this mechanism was evident in the name of his book: The Origin of Species, By Means Of Natural Selection. Natural selection holds that those living things that are stronger and more suited to the natural conditions of their habitats will survive in the struggle for life. For example, in a deer herd under the threat of attack by wild animals, those that can run faster will survive. Therefore, the deer herd will be comprised of faster and stronger individuals. However, unquestionably, this mechanism will not cause deer to evolve and transform themselves into another living species, for instance, horses. Therefore, the mechanism of natural selection has no evolutionary power. Darwin was also aware of this fact and had to state this in his book The Origin of Species: Natural selection can do nothing until favourable variations chance to occur.30

…So, how could these "favourable variations" occur? Darwin tried to answer this question from the standpoint of the primitive understanding of science in his age. According to the French biologist Lamarck, who lived before Darwin, living creatures passed on the traits they acquired during their lifetime to the next generation and these traits, accumulating from one generation to another, caused new species to be formed. For instance, according to Lamarck, giraffes evolved from antelopes; as they struggled to eat the leaves of high trees, their necks were extended from generation to generation. Darwin also gave similar examples, and in his book The Origin of Species, for instance, said that some bears going into water to find food transformed themselves into whales over time.31 However, the laws of inheritance discovered by Mendel and verified by the science of genetics that flourished in the 20th century, utterly demolished the legend that acquired traits were passed on to subsequent generations. Thus, natural selection fell out of favour as an evolutionary mechanism.

…In order to find a solution, Darwinists advanced the "Modern Synthetic Theory", or as it is more commonly known, Neo-Darwinism, at the end of the 1930's. Neo-Darwinism added mutations, which are distortions formed in the genes of living beings because of external factors such as radiation or replication errors, as the "cause of favourable variations" in addition to natural mutation. Today, the model that stands for evolution in the world is Neo-Darwinism.

Source: The Scientific Collapse of Darwinism © 2005, Harun Yahya

Neo-Darwinists suppose that genetic mutations within certain species can account for speciation, yet, while genetic mutations do occur, the kind that are beneficial have only been found at the molecular level, accounting for things like resistance to certain diseases. For Darwin's macroevolution theory to be true, the mutations necessary to change one animal into another would have to affect the animal's morphology (the shape and structure of its body). To date there has been no evidence that beneficial mutations affecting morphology have occurred in the wild. Mutations of this sort have proved to be either benign or detrimental to the survival of the mutant, utterly contradicting the Darwinist view.

Source: The Problem With Evolution, Intolerance For Independent Thought © 1997-2005, Ether Zone. Edward L. Daley

6.37   NON-DISCLOSURE AGREEMENT

A confidentiality agreement not to divulge information relating to a software project, to anyone outside that project. The agreement would normally be used to stop the disclosure of original inventions. And it may sometimes be incorporated into the contracts of the staff involved in that project. But the agreement has often been abused.

The Non-Disclosure Agreement or NDA has often been reinforced by notices on the documents of the project, proclaiming each one as 'Strictly confidential'. In the Computer Games industry, these documents have included the game design and technical design. Even though such notices have been redundant.

Some in the Games industry use the agreement for marketing purposes. Since they believe that a large amount of sales of a game would be lost without the use of such agreements, to stop the general theme or licence for a game being made public, before it was released.

An example of what they would consider a serious violation of an NDA would be for staff to allow some visitors, to a company, to catch a glimpse of a game based on a comic book hero, such as Batman, while this was being developed. Not for a glimpse of any original character. Nor for any technical detail about the game. Nor for any in-depth knowledge about the content, the story or the game design. But merely divulging the knowledge of the game being based on Batman would be considered, by them, to be a serious violation. And any staff at the company, deemed responsible, would be given some kind of formal warning, if not dismissed, for even revealing this to someone outside of a project.

Nevertheless, this would be an abuse of an NDA. Non-Disclosure Agreements were historically used for technical innovations in the Software Industry: not for marketing strategies. And such a use would be an abuse. It would not matter whether the financial backers of a project required the Software Developer to keep the project secret. Some believe that this would excuse the abuse of NDAs, by Software Developers, in the Computer Games Industry.

But the abuse of NDAs, by their financial backers, would in no way excuse the abuse of NDAs by the Software Developers: the two abuses would not cancel each other out. 'Two wrongs do not make a right!' as the proverb says.

Often interviewees will have to sign an NDA before being interviewed by a Software Developer in the Computer Games industry. In case the interviewee is shown or given a copy of a game that has yet to be publicly released. The NDA would include the same kind of clauses you would find in software licences that accompany games that were released. That is to say, you will find clauses that deny any warranty of the software for any purpose, to the software user. These clauses, just like the clauses in software licences, are further evidence of how the software user is denied a role in the Software Evolution Process used to build the game. Even though the Software Evolution Process is nominally meant to be based on feedback from the software user. Furthermore, these clauses undermine the implication that the NDA protects something that is valuable to the Software Developer. How can the software be valuable if it is not fit for any purpose? And the Software Developer denies any warranty of the software for any purpose? Unless the Software Developer is saying it is valuable strictly in the Marketing sense and sales sense. That is to say the ability of the Developer to either market itself or its products or to increase its sales or share value only. But it is not valuable in the practical application sense or usefulness to society sense. There is an example of a NDA in Figure 6.1.

DIRECTORY: ROUND_SCORE

The user's current score for the round. Can be followed by an optional ROUND_COMMENT comment

Cued by:
- The finishing of the hole.

Notes:
- Can be used for single day tournaments
- Can be used for single rounds (not tournaments)

SUB-DIRECTORY: 20OVER_ROUND_SCORE
Golfer on 20 over par for the round

20 over par for the round
20 over par for the day here.
20 over.
20 over for the day.
20 over par.

SUB-DIRECTORY: 19OVER_ROUND_SCORE
Golfer on 19 over par for the round

19 over par for the round
19 over par for the day here.
19 over.
19 over for the day.
19 over par.

SUB-DIRECTORY: 18OVER_ROUND_SCORE
Golfer on 15 over par for the round

18 over par for the round
18 over par for the day here.
18 over.
18 over for the day.
18 over par.

SUB-DIRECTORY: 17OVER_ROUND_SCORE
Golfer on 17 over par for the round

17 over par for the round
17 over par for the day here.
17 over.
17 over for the day.
17 over par.

SUB-DIRECTORY: 16OVER_ROUND_SCORE
Golfer on 16 over par for the round

16 over par for the round

FIGURE 6.1 A typical example of a page from an incomplete game design in a Software Evolution Process of Slippery Games Inc. (Source: A typical game design from a Software Evolution Process of Slippery Games Inc. showing the excessive use of confidentiality to shroud in mystery unoriginal game designs. 2005. Anonymous.)

'NON-DISCLOSURE AGREEMENT

THIS AGREEMENT (the "Agreement") is made between

and

Candidate name:

Address:

and entered into as of date:

...

5. *All Confidential Information and Confidential Materials are and shall remain the property of Disclosing Party. By disclosing information to Receiving Party, Disclosing Party does not grant any express or implied right to Receiving Party to or under Disclosing Party patents, copyrights, trademarks, or trade secret information. Receiving Party may only use such Confidential Information and Confidential Materials to evaluate*
6. *Receiving Party shall return all originals, copies, reproductions and summaries of Confidential Information or Confidential Materials at Disclosing Party's request.*
7. *If Disclosing Party provides pre-release software to Receiving Party, such prerelease software is provided "as is" without warranty of any kind. Receiving Party agrees that neither Disclosing Party nor its suppliers shall be liable for any damages other then by intend whatsoever relating to Receiving Party's use of such pre-release software.'*

Source: A typical Non-Disclosure Agreement for interviewees of Slipper Games Inc. 2024. Anonymous

6.38   ORIGINAL GAMES (RELEASED EACH YEAR)

The majority of the computer games that have been released each year have tended to be clones or sequels of successful games from the past. Or these are based on popular franchises from other industries such as characters from comic books, book series, films, TV series, sports or toys. The top ten most successful games released each year rarely include an original game. This is not only a reflection of the familiarity and confidence in these brands, but the unoriginality of the alternatives.

Jason Della Rocca, executive director of the International Game Developer's Association, or IGDA, said he hopes the industry will be able to offer more diversity to its fans and new consumers...

That leads to other issues in the industry, which is seeing more risk aversion, Della Rocca said.

"They know they can bank on the success of previous games," Della Rocca said. "There's a lack of innovation, originality, a lot of sequels, a lot of games based on movies, book or comic book licenses. They don't want to risk creating their own worlds."

Source: The evolution of video games ... to now, where games evolve alongside the consoles © 2005, The State News. Michigan State University. Lauren Phillips

6.39   MYTHICAL MAN MONTH

The theory is if it took one person a certain amount of time to complete a task, it would take two approximately half the time to complete that same task.

The theory of a Man Month was devised to plan and control projects. It was a unit devised to quantify how long it took one member of staff to complete a task, in months. So that the leadership of the staff could use this to methodically project how much time it should take other members of staff, especially a larger group of staff, to complete the same task.

But such projections were not methodical. Since these did not take into account the added complexity of the task caused by introducing more staff. For example, the dependencies between the staff would require more and clearer communication between them as their numbers increased. The dependencies between the staff would mean that the more new staff that were introduced, the more time would be required of the old staff, to teach the new staff about the task, before they could contribute. And the more errors the new staff would make in the meantime would also consume more time of the old staff. The dependencies between the staff would also require greater coordination between them for access to limited resources.

It was partially an exasperation with the belief in the Man Month that Frederick P. Brooks was inspired to write his book which made the phrase Mythical Man Month famous. He was working at the International Business Machines Corporation (IBM) at the time when he noticed the use of the Man Month by the leadership of the staff.

Adding manpower to a late software project makes it later.

Source: The Mythical Man Month: Essays on Software Engineering,
20th Anniversary Edition © 1995, Addison-Wesley. Frederick P. Brooks

6.40   BUG

A software error. The name comes from an anecdotal story about an error caused by a moth short-circuiting an old computer.

6.41   BUG DATABASE

A Database of the errors found in a software product. In the Computer Games industry, each error recorded in the Database would have several properties.

The first of these properties would be a number used to identify the error. This would normally correspond to the order in which the error was discovered (e.g. 0001 for the first, 0002 for the second, 0003 for the third etc.)

A second property would be a brief summary. This would describe where the error appeared in the game and what happened just before it appeared.

A third property would be a classification of its severity. Each error would be classified by a letter ranging, from 'C' for the least severe, to 'A' for the most severe.

A fourth property would be a classification of the component of the game design the error was related. This would be in the form of one of several keywords.

This could be either 'Crash', for errors which literally stopped the game. Or this could be 'Text', for errors on the menus and the wording that appeared somewhere in the Game World. Or this could be 'Function', for errors relating to the order in which some events occurred in the game. Or this could be 'Graphics', for errors in the various 2D images and 3D models. Or this could be 'AI', for errors in the behaviour of computer-controlled characters. Or this could be 'Logic', for errors which just appeared counterintuitive, or contradicted other features of the game. Or this could be 'Audio', for errors relating to the sound and music heard during the game.

A fifth property, which every error would have, would be the name of the member of staff assigned to fix it. And this would be accompanied by some kind of status indicating whether the error was not fixed, has been claimed fixed but not verified or been fixed and verified.

However, the Database containing all of these properties would be used as a substitute for a complete game design. So many of the properties recorded would, at best, be misleading and at worst false.

Firstly, the total number of errors in the Database would bear no relation to the total number of errors in the product. Since there never was a complete game design to compare the product to. Thus, the order in which the errors would be discovered would be arbitrary. And the identity number of these errors would be random. For example, all of the errors that were discovered in one area of the game e.g. the Front End menus, would not have consecutive identifying numbers. Nor would these be the lowest numbered errors in the game e.g. 0001, 0002, 0003 etc. Instead the lowest numbered errors would merely reflect the first superficial errors discovered in the game. Not even the most significant errors.

Secondly, the quality of the summary, of the errors in the product, would vary from entry to entry. Some of the entries would be produced by the Game Testers, of the Software Developer. Others would be produced by the Game Testers, of their financial backers. Some would be produced by Game Producers, Game Designers, Game Artists and Sound Designers. And others would be produced by Game Programmers. The quality of each of their entries would vary greatly, depending on their knowledge of the history of the project, and their intuitions. As they would all have no authoritative game design to refer to.

Thirdly, the severity of each error would likewise be subject to the whims of their intuitions. All errors that caused the game to literally stop would be classified as 'A'. And those missing features which were necessary to meet any standards set by an important third party, such as the financial backers, or game console manufacturers, would also be classified as 'A'.

However, features which contradicted what little of the game design that had been written, at the beginning of the production process, may or may not be classified as 'A'. Even though all of these would be factual errors that could be confirmed by documentation.

Features which some of the testers just found very annoying, but did not stop the game, would also be classified as 'A'. While other incongruous features, they would merely classify as 'B' or 'C' even though, by their own admissions, these would all be errors.

It would all depend on either whether those who saw these errors knew about the original design. Or it would depend on what they could speculate about that design, from the previous versions of the game they had seen. Or it would depend on the informal discussions, which they had conducted with other members of staff, to disperse their ignorance about the design.

The different levels of severity, assigned to the errors in the Database, would merely be there to compensate for this ignorance. Furthermore, these levels would even include one called 'suggestions'. That is to say, new ideas which they feel should have been part of the original game design. And some of these 'suggestions' would eventually end up being promoted and reclassified as 'A', 'B' or 'C'.

Fourthly, the keywords corresponding to the different components of the game design, assigned to each error, would imply that the game design was complete. Since only if all the components of the game design were known could all the categories of errors in these components be known. But, of course, this would be false. The game design would not be complete. And those deciding the different keywords would be uncertain about how to classify certain errors. Some of their classifications would overlap with others, because of this uncertainty. Thus, their classifications would be almost meaningless.

Take, for example, the set of keywords described earlier, one of which included 'Design'. Most of this set would be completely redundant since all errors in the product would be the direct result of the incomplete game design and could be classified as 'Design'.

Finally, the status of each error, recorded in the Database, would be just as meaningless. Since those verifying the errors that were fixed would be suffering from the same ignorance, about the game design, as those who discovered these errors in the first place. Their uncertainty would be reflected in many ways. One of these would be their use of adjectives as nouns.

For example, they would use a word like 'Shippable', to categorise the status of an error. This would refer to an error that some abortive attempt had been made to fix. And the error was consequently decided not to be severe enough to stop the release of the product.

Another instance of the lack of credibility in the Database would be amongst the other options which the software itself provides. For the member of staff assigned to fix an error would be given the option to close its entry in the Database and have it waived. Provided, that is, this was accompanied by some explanation. However, frequently, this explanation would be rejected by the tester who made that entry, who would reopen it, changing its status back again. So the error would go back and forth, in this manner, for several days, between the two states because of the absence of any authoritative device for settling disputes, including the Bug Database.

Further examples of the meaninglessness of the classification of Bugs in the Bug Database would be evident by the questions the staff frequently ask, openly, at the end of production.

How did this go on for so long unnoticed?

This would be asked when it became apparent, at a very late stage, that some feature of the product was erroneous. The implication would be that this feature was so obviously erroneous that it was incredible it managed to escape anyone's notice, including those who made entries in the Bug Database. But this implication would be false. The feature was in all likelihood noticed, but ignored. All due to the lack of any authoritative device that would naturally draw attention to its flaws, including those who made entries in the Bug Database.

Software Developers frequently use a Bug Database to monitor ad hoc processes, such as the Software Evolution Process. Like that process, it never reaches any satisfactory conclusions. And it would be rare for all the entries in the Database to be resolved. Since the size of the Database has only an incidental bearing on the quality of their final product, different Developers have different thresholds for the number of unresolved entries, below which the product would be released to the public. Some have a threshold of around 50, while others have a threshold of around 100. Others have a threshold even higher and have no qualms referring to the unresolved entries euphemistically, as 'issues'. There is an example of a graph showing the total new errors reported in a Bug Database in Figure 6.2. There is an example of a Web Page showing the total number of Bugs in different categories in Figure 6.3.

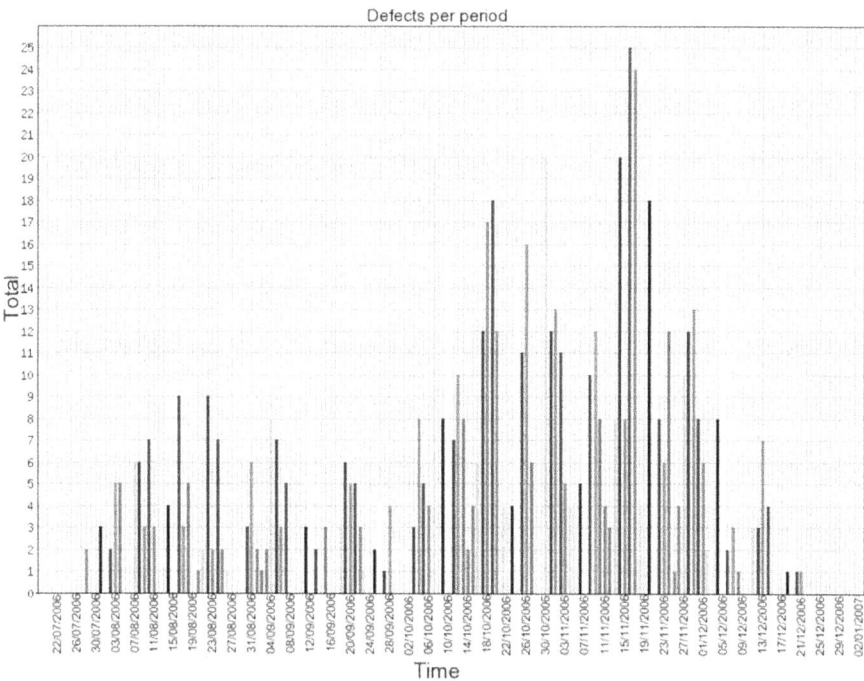

FIGURE 6.2   A typical example of a bar graph showing the new errors reported in a Bug Database, each day, over the course of 6 months, during a Software Evolution Process of Slippery Games Inc. (Source: A typical report of new errors in a Bug Database, during 6 months of the final, testing phase of a Software Evolution Process of Slippery Games Inc. Anonymous. January 2007.)

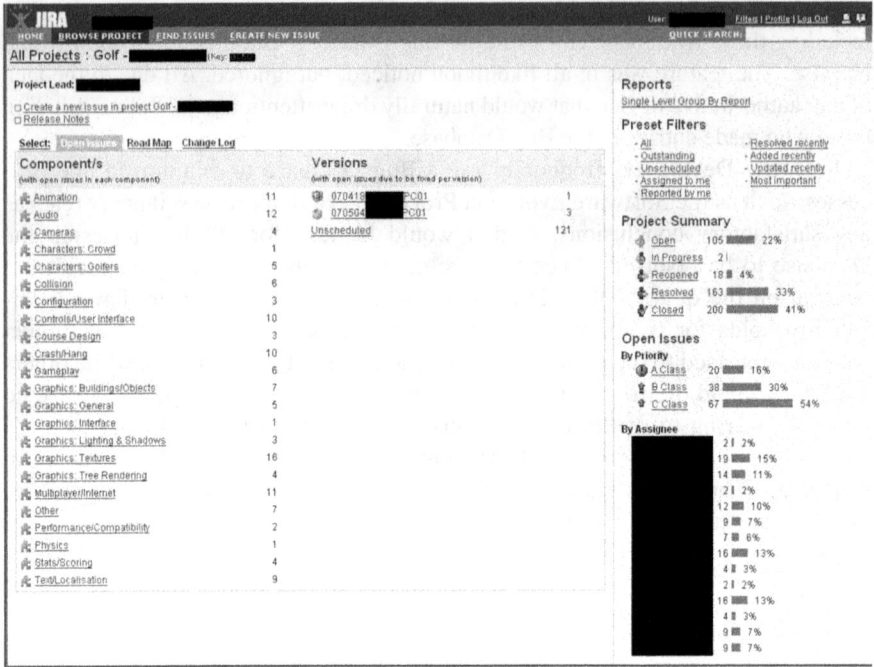

FIGURE 6.3 A typical example of a Web Page showing the high number of Bugs reported in a Bug Database during the QA phase, of a Software Evolution Process of Slippery Games Inc. (Source: A typical report of errors in a Bug Database, during the final, testing phase of a Software Evolution Process of Slippery Games Inc. Anonymous. June 2007.)

6.42 MEMORANDUM

The primary source of the explanation of the tools being used, in a Software Evolution Process, are memoranda. These would usually be delivered through electronic mail (e-mail).

Fyi, we sometimes have poorly authored Logan strings with missing token indices, and it's important to understand how these should be handled when adding strings into the code. There may have been emails sent around about this in the past, but I can find them - so here's a new example.

A made up news item with 3 alternative phrases:

String _ Id.sch#1 ""#4-Number# year old #2-Player# has suffered #3-Injury#.""

String _ Id.sch#2 ""#4-Number# year old #2-Player# was injured in the fixture against #7-Club#.""

String _ Id.sch#3 ""#2-Player# has suffered #3-Injury#. Manager #6-Staff# says he'll be back playing for #8-Club# in no time.""

The set of tokens used across all the alternatives is as follows:

2-Player

```
3-Injury
4-Number
6-Staff
7-Club (other club)
8-Club (player's club)
```

Not all token indices have been used (there's no token 1 or 5). Part of the text exporter process will recognise this and will renumber the tokens, e.g. as if the strings had been authored as follows.

```
String _ Id.sch#1 ""#3-Number# year old #1-Player# has suffered
    #2-Injury#.""
String _ Id.sch#3 ""#3-Number# year old #1-Player# was injured
    in the fixture against #5-Club#.""
String _ Id.sch#2 ""#1-Player# has suffered #2-Injury#. Manager
    #4-Staff# says he'll be back playing for #6-Club# in no time.""
```

However, the simplest way to think of this (rather than how tokens are renumbered) is that the token order specified in a call to NMAddNewsItem needs to match the numeric order in the string. So to add this news item you might implement it as follows (assuming a headline containing just the Player token):

```
NMSetParams8 (NM _ RESET, ""tytmStClClNuPlIn"", tNewsItem::NEWS _
    TYPE _ XXXX, Club,
NIStaff(Manager),
NIClub(Club),
NIClub(OtherClub),
etc.
NMAddNewsItem (""Pl.PlInNuStClCl"", NEWS::THE _ HEADLINE _ FOR _
    STRING _ ID <news::THE _ HEADLINE _ FOR _ STRING _ ID>, NEWS::
    STRING _ ID <news::STRING _ ID>) ;
```

Bear in mind that strings entered into DevStrings.cpp won't have been renumbered (as it's one of the exporter tools that does this). This means that the above code would produce a broken news item until after a text export. This isn't really a problem so long as you're aware of it (and most of our strings are correctly authored and so don't have this issue).

If you have a poorly indexed string that you'd like to display correctly while it's still in DevStrings, I'd suggest manually renumbering it in DevStrings. If the string has alternative phrases but is already in Logan, you only really need to copy a single phrase into DevStrings and edit that.

> *Source: A typical E-mail from a Software Evolution Process of*
> *Slippery Games Inc. Anonymous. June 2006*

'This README provides instructions for doing a [Redacted] text export.

General Notes

=============

Do not have [Redacted] project open in Visual Studio.

The text exporter tool project is under [Redacted]/Tools/XXXXXGames/ Apps/TextToDBConverter. This needs to have been built (do a release build) beforehand (but doesn't need to be rebuild each time). (See instructions below)

The tool scans the source code (under/[Redacted]/[Redacted]) and. xml files (under/[Redacted]/PS2/Screens/[Redacted]/XBox/Screens/ [Redacted]/PSP/Screens) to find out which strings are actually used. Only these strings are exported. If you want to export strings that aren't used yet, put the string IDs (enums) in code either in a comment in the form ##string_id## or in a #if 0 block.

Checking Out Files

==================

- *Sync up on everything (i.e. get latest in Alienbrain)*
- *Open up [Redacted]/Data/Logan/Exports in windows explorer*
- *Run CheckoutFiles.bat/CheckoutFiles_EnglishOnly.bat (depending on whether you are doing all languages or just English)*
- *Build the text exporter tool (see above) (make the entire TextToDBConverter folder writable before compiling)*
- *Check the file history of e.g. [Redacted]/BTR/Data/News.h to see the version number of the last export.*
- *Check that you've gained a change set in Alienbrain and rename to append the new version number e.g. "Text export (English only) v25"*
- *Note that there are now change sets in both [Redacted] and [Redacted]*

Text Data Conversion

====================

- *ConvertAll.bat/ConvertAll_EnglishOnly.bat*
- *CopyAll.bat/CopyEnglish.bat and check that the copy commands each generate a '1 files copied' result*
- *CopyHeaders.bat and again check for copy command errors*
- *Copy the entire [Redacted]\CD\PS2\Lang folder to the platform you are testing (e.g. PC = [Redacted]\CD\PC\Lang) but do not copy the Japanese folder or Lang.dat*

Rebuild

=======

- *Open up [Redacted] in Visual Studio*
- *Open TextFile/DevStrings.cpp and remove any new strings people have added. These will usually have been adding into Logan as well and so should come through in the export.*
- *Do a full rebuild*
- *It may be necessary to re-instate some string enums and their matching entries in DevStrings.cpp. This should be pretty straightforward by checking the file changes in Alienbrain.*

Test

====

- *XBox only: Run XBCP or rebuild a single file and relink to cause the post-build steps to be run. Probably not required if rebuilding-all using Incredibuild, as running (F5) tends to cause VC to rebuild a few files and generate the browse info file.*

- *Run up the game.*
- *Any assertions talking of text file versions means the version in the headers and the data files don't match. Check files have been copied properly.*
- *Submit the pending changes for your text export change set.'*

Source: A typical Memorandum from a Software Evolution Process of Slippery Games Inc. Anonymous. 2006

So for example if we had a loop that was combining data from two arrays and storing a result eg:

sint32 a[1000], b[1000], c[1000];

for (i = 0; i < 1000; i++)

{

c[i] = a[i]+b[i]

}

If we are not careful this algorithm could result in a considerable number of cache misses. The worst case being when the a,b and c arrays are all 4k aligned:

i =0

a[0] read into set 0 line 0x000 - 128byte read

b[0] read into set 1 line 0x000 - 128byte read

c[0] stored into set 0 line 0x000 and dirty bit set for cache line -

But before this occurs the 128 bytes at c[0] are read into the cache -

128byte read. note this set 0 is the least recently used

i = 1

a[1] read into set 1 line 0x000 - 128byte read (set 1 least recently used)

b[1] read into set 0 line 0x000 - 128byte read - But before this read

the cache line is placed in the write buffer since dirty bit set

c[0] written out of write buffer - 128byte write

c[1] stored into set 1 line 0x000 and dirty bit set for cache line -

But before this occurs the cache line is read into the cache - 128byte

read.

etc.

As you can see this results in 4 128 byte memory accesses (3r and 1w) per iteration of the loop. To ensure that you are not hit by this you have to make sure that a b and c are at least 128 bytes apart. Although since we have 2 way set associative cache one of a n and c can have a similar alignment to the other.

So for example with a and b being 2k aligned (remember to use the size of cache divided by n for a n way associative cache) and c being aligned 1k into a 2k aligned block we would have the following:

i =0

a[0] read into set 0 line 0x000 - 128byte read

b[0] read into set 1 line 0x000 - 128byte read

c[0] stored into set 0 line 0x400 and dirty bit set for cache line -

But before this occurs the 128 bytes at c[0] are read into the cache -

128byte read.

i = 1

a[1] read from set 0 line 0x000 - cache hit

b[1] read from set1 line 0x000 - cache hit

c[1] stored into set 0 line 0x400 - cache hit.

...

i = 32

a[32] read into set 0 line 0x080 - 128byte read

b[32] read into set 1 line 0x080 - 128byte read

c[32] stored into set 0 line 0x480 and dirty bit set for cache line -

But before this occurs the 128 bytes at c[32] are read into the cache

```
- 128byte read.

etc.
```

So now the algorithm does 3 128 byte accesses for every 32 itera-
tions until the cache the values being added by writing to c[0-256]
need to be written out of the cache due to a[256-384] and b[256-386]
being read into the cache.
*   If you take the psp with its 61 cycles for the read this optimiza-*
tion speeds the code up by 2000%!

Source: A typical E-mail from a Software Evolution Process of
Slippery Games Inc. Anonymous. March 2006

6.43   SOFTWARE ARCHITECTURE

A description of a system for producing software. It includes a description of the components of the system, the relationship between these components and the principles that
govern how these components change. The components may be as large as a software
library or as small as a single software module. The components can also vary from any
software documentation to any software tool or member of staff required by the system.
A software architecture can serve as a basis for a software design or (since all the components do not have to be software components) a software production process.

The software architecture of a program or computing system is the structure or struc-
tures of the system, which comprise software components, externally visible proper-
ties of those components, and the relationships among them.
*   …the architecture embodies information about how the components interact with*
each other. This means the architecture specifically omits content information about
components that does not pertain to their interaction.
*   …the definition does not specify what architectural components and relationships*
are. Is a software component an object? A process? A library? A Database? A com-
mercial product? It can be any of these things and more.
*   …the behaviour of each component is part of the architecture, insofar as that*
behaviour can be observed or discerned from the point of view of another component.
This behaviour is what allows components to interact with each other, which is clearly
part of the architecture. Hence, most of the box-and-line drawings passed off as archi-
tecture are in fact not architectures at all. They are simply box-and-line drawings.

Source: Software Architecture in Practice © 1997, Addison-Wesley.
Bass, Clements and Kazman

The structure of the components of a program/system, their interrelationships, and
principles and guidelines governing their design and evolution over time.

Source: IEEE Transactions on Software Engineering © 1995, Institute of Electrical
and Electronics Engineers. David Garlan and Dewayne Perry

6.44   LOGIC BRANCH

The point where a software procedure decides to follow one path or another, in its
overall task.

6.45   ID

Identifier. A word which identifies one or more sets of data. In a Database, the ID of each Record is a special Field known as the Key Field or Primary Key.

6.46   SOFTWARE PROCEDURE

A sequence of instructions, for a computer, to perform a task. The sequence can be used again and again to repeat that task.

6.47   DATABASE

A collection of data arranged for ease and speed of search and retrieval.

6.48   RELATIONAL DATABASE

A Database where all the software data, and the relationship between these, are organised in tables with rows, known as Database Records, and columns, known as Database Fields.

> ...(RDBMS - relational database management system) A database based on the relational model developed by E.F. Codd. A relational database allows the definition of data structures, storage and retrieval operations and integrity constraints. In such a database the data and relations between them are organised in tables. A table is a collection of rows or records and each row in a table contains the same fields. Certain fields may be designated as keys, which means that searches for specific values of that field will use indexing to speed them up...
>
> Source: RDBMS. The Free On-line Dictionary of Computing
> © 1993-2001, Denis Howe

6.49   DATABASE TABLE

A collection of Database Records. A group of related data about entities (e.g. characters, locations or items in a Game World) which share the same properties in a Relational Database.

6.50   DATABASE RECORD

A collection of Database Fields. A group of related data about a single entity in a Relational Database (e.g. the name of a character in the Game World, its location, its health, its inventory).

6.51   DATABASE FIELD

A single property of an entity in the Database (e.g. the name of a character in the Game World). An element in a Database Record.

6.52 PRIMARY KEY

The first Database Field of a Database Record that is used to identify that Record, search for it and refer to it. This has to be a unique word or number.

6.53 DATABASE ADMINISTRATOR

A company employee responsible for the design and management of one or more Databases. The employee is also responsible for the evaluation, selection and implementation of the Database management system.

6.54 OPEN DATA FORMAT

The description of the layout of data in a Database and how each data is used. This description is freely available for all software applications to use to read and modify the Database.

6.55 GAME SOFTWARE

The game modules and game engine that make up a computer game.

6.56 GAME CONTROLLER

A device used to control the User Interface, including the player and other characters, of a game.

6.57 SOUND STREAM

A recorded sample of sound encoded in a special data format.

6.58 GAME TIME

The number of seconds since a game was started.

6.59 UNIT OF GAME TIME

The assumed minimum time between successive updates of a Game World.

In the Event-Database Architecture, this is the minimum time between successive updates of a Host Module. The real time may exceed this limit, because the total time it takes to update all of the Host Modules may be too long.

Most modern Computer Games are dependent on graphics and the rate at which the graphics or Frames of the Game World are displayed. So the unit of game time is the minimum time between successive updates of the Frames. Since this depends on how many graphics are displayed in each Frame, the rate at which the Frames are displayed changes depending on where the player is in the Game World. And how many items are in that part of the World? Thus, the unit of game time changes and is not fixed. And the game modules have to take this into account when updating items in the Game World e.g. the game modules that update the physics of the Game World.

6.60   POLYGON

A closed plane shape, with three or more sides. Triangles, sometimes called 'Tristrips' in the degenerative language of the Software Evolution Process in the Computer Games industry, are used to make up a 3D model. Quadrilaterals, sometimes called 'Quads' in the degenerative language, are used to mark the position of a rectangular 2D image.

6.61   VERTICES

The point at which two or more sides of a shape meet. Three vertices are used to form triangles, which make up a 3D model. Four vertices are used to form a quadrilateral, which marks the position of a rectangular 2D image.

6.62   VECTOR

The magnitude and direction of a physical quantity e.g. force, speed etc.

6.63   NORMAL VECTOR

A Vector that is perpendicular to the side of a 2D shape or plane of a 3D polygon which simply specifies the direction in which a 2D or 3D surface is facing.

In 2D geometry, a Normal Vector may be used to specify which direction the sides of a polygon are facing. Each Vector would be perpendicular to one side of the polygon, pointing away from its centre.

In 3D geometry, a Normal Vector may be used to specify which direction the surface of a polygon was facing. If the polygon were isolated, then it would have one Normal Vector, that was perpendicular to its plane. However, if the polygon were part of a 3D model, then each vertex of the polygon may have a Normal Vector. The choice between the two methods would depend on how well you wanted to describe the surface of the model. The latter method would allow you to describe how each polygon connects to any adjoining polygon. The Normal Vector at each vertex would describe the combined surface of the two adjoining polygons that share that vertex.

6.64   TEXTURE

A 2D image which is used to fill in a polygon. Only the region of the image specified by the Texture coordinates, of the polygon, is used to fill it in.

6.65   TEXTURE COORDINATES

A set of points describing the region of an image which should be used to fill in a polygon. There are the same number of points as there are vertices in the polygon. Each point corresponds to one, unique vertex.

6.66   FRAME

A single image in an animated sequence. A single image of an animated world.

6.67   X POSITION

The position of a body along the X-axis in a 2D or 3D space.

6.68   Y POSITION

The position of a body along the Y-axis in a 2D or 3D space.

6.69   Z POSITION

The position of a body along the Z-axis in a 2D or 3D space.

6.70   X ANGULAR POSITION

The rotation of a body, in a local 2D or 3D space with an origin at its centre of mass, around the X-axis, in a plane perpendicular to the axis or the ZY plane.

6.71   Y ANGULAR POSITION

The rotation of a body, in a local 2D or 3D space with an origin at its centre of mass, around the Y-axis, in a plane perpendicular to the axis or the ZX plane.

6.72   Z ANGULAR POSITION

The rotation of a body, in a local 2D or 3D space with an origin at its centre of mass, around the Z-axis, in a plane perpendicular to the axis or the XY plane. In 2D space, the Z-axis does not exist and it's just an imaginary axis extending out from the 2D plane.

6.73   SOFTWARE RENDERING

Rendering items in 2D or 3D space using a Central Processor and main memory in a computer system. The Central Processor is sometimes called 'CPU'.

6.74   HARDWARE RENDERING

Rendering items in 2D or 3D space using a specialised Graphics Processor and Graphics memory in a computer system. The Graphics Processor is sometimes called 'GPU'.

The rendering is done by executing the same set of steps, during each Unit of game time or between each Frame of the 2D or 3D space being displayed on the screen. Each cycle through these steps is also called a 'rendering pass'. A computer game may use several cycles or 'rendering passes' between each Frame. In these cases, the results of one cycle or 'pass' is fed into the next cycle or 'pass'. To progressively build up the final image seen on the screen.

The names of these 'passes', and the terminology used to describe them, can be very cryptic because the steps involved use machine code and mathematics. But the branch of mathematics involved is basically Linear Algebra, Vectors and Matrix theory. And all these steps do in the end is render an image of a 2D or 3D Space of a Game World on a 2D screen.

In this chapter you'll learn about:

- What is a *rendering pass*
- Over 20 kinds of passes in Unreal – lighting, the base pass or the mysterious HZB
- What affects their cost (as seen in the GPU Visualizer)
- How to optimize each rendering pass

...

HZB (Setup Mips)

Responsible for:

- Generating the Hierarchical Z-Buffer

Cost affected by:

- Rendering resolution

The HZB is used by an occlusion culling method 1 and by screen-space techniques for ambient occlusion and reflections 2.

Source: Unreal's Rendering Passes © 2019. Oskar Świerad

6.75   NEAR AND FAR FOCAL LENGTH

The closest and furthest distance of the visible area or volume in front of a camera.

6.76   FIELD OF VIEW

The angle between the left-hand side and the right-hand side of the visible area or volume in front of a camera.

6.77   X SPEED

The speed of a body along the X-axis in 2D or 3D space.

6.78   Y SPEED

The speed of a body along the Y-axis in 2D or 3D space.

6.79   Z SPEED

The speed of a body along the Z-axis in 2D or 3D space.

6.80   X ACCELERATION

The acceleration of a body along the X-axis in 2D or 3D space.

6.81   Y ACCELERATION

The acceleration of a body along the Y-axis in 2D or 3D space.

6.82   Z ACCELERATION

The acceleration of a body along the Z-axis in 3D space.

6.83   X ANGULAR SPEED

The rotational speed of a body around the X-axis in 3D space.

6.84   Y ANGULAR SPEED

The rotational speed of a body around the Y-axis in 3D space.

6.85   Z ANGULAR SPEED

The rotational speed of a body around the Z-axis in 2D or 3D space.

6.86   X ANGULAR ACCELERATION

The rotational acceleration of a body around the X-axis in 3D space.

6.87   Y ANGULAR ACCELERATION

The rotational acceleration of a body around the Y-axis in 3D space.

6.88   Z ANGULAR ACCELERATION

The rotational acceleration of a body around the Z-axis in 2D or 3D space.

6.89   SOUND CHANNEL

A component of computer-generated sound, which can play back a sound (given the sound envelope, i.e. the shape of a sound wave or a sound stream) independently, or mixed with other sound channels.

6.90   ANALOGUE DEVICE

A device which produces data that measures a continuously variable, physical quantity e.g. the rotation of a Joystick about its X, Y or Z axes, the pressure applied to a button. These devices are used to control characters or User Interfaces in a game.

6.91   DIGITAL DEVICE

A device which produces data that measures a binary, physical quantity e.g. a joystick being moved to the left or right, a button being pressed or released. These devices are used to control characters or User Interfaces in a game.

6.92   OPERATING SYSTEM

A software that controls how other software shares resources on the same computer hardware.

6.93   SOFTWARE APPLICATION

A software programme that is used directly by a Software User, through a User Interface, to solve a problem.

When it is started by an Operating System which in turn is running on computer hardware, then it shares the resources on that hardware with other Applications. When it is started on its own on computer hardware without an Operating System, then it has exclusive access to the resources on that hardware.

6.94   PROCESS (IN AN OPERATING SYSTEM)

A software application or programme or routine that is running in its own space in computer memory. When it is started by an Operating System or another software application running on the same computer hardware, that System or Application can temporarily or permanently interrupt it. To allow other Processes to share the resources on the computer hardware. Before the Process is resumed.

6.95   THREAD (IN AN OPERATING SYSTEM)

A sub-process generated from another Process being run by an Operating System, which shares the same space in computer memory as its parent. This simplifies and speeds up the communication between the child and its parent Process. Unlike a Process which runs in its own space in computer memory, a Thread has to explicitly lock shared resources, and explicitly wait for resources to be released by other Threads. Since it shares that space in memory with its parent and its siblings.

6.96   TCP/IP ADDRESS

Transmission Control Protocol or Internet Protocol is a protocol for communicating between two computers on a local computer network. The Address of each computer is a unique word, normally made up of 4 numbers separated by dots, used to identify that computer, and the source and destination of a message.

6.97 PORT NUMBER

A number that represents a channel through which messages can be sent or received by a computer on a network. Several messages may be sent or received in parallel on the different channels on the same computer.

6.98 USERNAME

The unique name of a Software User used to identify that User and the resources e.g. files, threads or processes, that they own in an Operating System or Software Application.

6.99 PASSWORD

The unique word that only a Software User knows and uses to authenticate their access to resources available on an Operating System or Software Application, that they own.

6.100 AUTHENTICATION TOKEN

A unique encrypted word that is generated by a computer, from a Username and Password, to authenticate that User's access to resources available on an Operating System or Software Application, that they own. Usually the token last for a limited time before having to be renewed by the System or Application.

6.101 RELATIONAL DATABASE MANAGEMENT SYSTEM

Software that creates, edits and queries a Relational Database. It normally includes a programming language, Structured Query Language or SQL, that allows you to query the database. The software is also responsible for making the Database appear as one system. Even though it may be distributed across several files, in several locations in a File System, across several Operating Systems and across several computers on a local computer network.

6.102 THE SOFTWARE ARCHITECTURE

The description of all the software modules (and staff) which would be required to completely implement an Event-Database Architecture. See the chapter entitled "The Software Architecture" in the book *Event-Database Architecture for Computer Games: Volume 1, Software Architecture and the Software Production Process.*

6.103 ENTITY-RELATIONSHIP DIAGRAM

A diagram which shows all the items (or entities) stored in a Relational Database and the relationship between these items.

6.104 BASIC SET THEORY

A branch of mathematics concerned with producing rational conclusions, about items in the real world, by abstracting these into groups or sets. Basic Set theory can be used to create a model of any computer software or hardware. Or a model of any component of software or hardware.

6.105 HASH-TABLE

A table of information where the entries have been positioned using a Hashing Function. A Hashing Function is a software procedure which tries to map any random set of numbers or words onto a non-overlapping set of numbers, within a limited range. This speeds up the reading and writing of entries in that table.

6.106 PROFESSIONAL DATABASE SOURCES

Database Systems: Concepts, Languages, Architectures © 1999, McGraw-Hill Education. Paulo Atzeni, Stefano Ceri, Stefano Parabosci and Riccardo Torlone.

6.107 DATA STRUCTURE SOURCES

Introduction to Algorithms, second edition © 2001, MIT Press. Thomas H. Cormen, Charles E. Leiserson, Ronald L. Rivest and Clifford Stein.

6.108 APPLIED MATHEMATICS SOURCES

Vectors in Two and Three Dimensions (Modular Mathematics S.) © 1995, Butterworth-Heinermann. Ann Hirst.

6.109 PHYSICS SOURCES

Computational Dynamics, second edition © 2001, Interscience. Ahmed A. Shabana.

6.110 MATHEMATICS SOURCES

The Geometry Toolbox for Graphics and Modeling © 1998, A. K. Peters. Gerald Farin and Dianne Hansford.

6.111 COMPUTER GRAPHICS SOURCES

Computer Graphics: Mathematical First Steps (c) 1998, Prentice Hall. Patricia Egerton and William Hall.

6.112   SOUND ENGINEERING SOURCES

The DSP Handbook © 2001, Prentice Hall. Andrew Bateman and Iain Paterson-Stephens.

6.113   FREE SOFTWARE

Software which can be freely copied, redistributed or modified according to its GNU Public Licence. The software comes with the computer files used to build it. So that the software can be easily modified.

Since the licence has required that it can be easily modified and redistributed, this has produced a de facto software architecture for software with this licence. Namely, it has defined a relationship between the Software Developer and the user, with regard to the software. The user (and by implication the Software Developer) has the right to copy the software. And all its components have had to be comprehensible and editable to a degree by the user (and by implication the Software Developer). And that in turn has added a software architecture, directing how the software has modified over time, on top of whatever explicit architecture it already had. So, although like almost all software with normal commercial licences, most Free Software too has been developed through a Software Evolution Process, this software architecture mitigates the otherwise harmful effects of the Process. And it makes Free Software more reliable and robust.

"Free Software" is a matter of liberty, not price. To understand the concept, you should think of "free" as in "free speech", not as in "free beer".

Free software is a matter of the users' freedom to run, copy, distribute, study, change and improve the software. More precisely, it refers to four kinds of freedom, for the users of the software:

- The freedom to run the program, for any purpose (freedom 0).
- The freedom to study how the program works, and adapt it to your needs (freedom 1). Access to the source code is a precondition for this.
- The freedom to redistribute copies so you can help your neighbor (freedom 2).
- The freedom to improve the program, and release your improvements to the public, so that the whole community benefits (freedom 3). Access to the source code is a precondition for this.

Source: What is Free Software? © 2008, Free Software Foundation Inc.

6.114   ELECTRONIC DOCUMENTATION SOURCES

OpenOffice.org 1.0 Resource Kit © 2003, Prentice Hall PTR. Solveig Haugland and Floyd Jones.

6.115   PROGRAMMING SOURCES

The C Programming Language, second edition © 1988, Prentice-Hall. Brian W. Kernighan and Dennis M. Ritchie. ISBN 0-13-110362.

The C++ Programming Language © 1993, Addison-Wesley. B. Stroustrup.

6.116   COMPILER SOURCES

Using GCC: The GNU Compiler Collection Reference Manual for GCC 3.3.1 © 2003, Free Software Foundation. Richard M. Stallman and the GCC Developer community.

6.117   REVISION CONTROL SOFTWARE

A tool used to store, retrieve, log, identify and merge different versions of software in production. It stores all the sources which produced each version e.g. the documentation of the software designs, the computer files used to build the software modules, the software data etc.

6.118   REVISION CONTROL SOURCES

Essential CVS © 2003, O'Reilly and Associates. Jennifer Vesperman.

Applying RCS and SCCS: From source control to Project Control (Nutshell Handbook) © 1995, O'Reilly and Associates. Don Bolinger and Tan Bronson.

6.119   FILE LIBRARY SOURCES

The Standard C library © 1991, Prentice Hall PTR. P. J. Plauger.

6.120   GRAPHICS LIBRARY SOURCES

OpenGL Reference Manual: The Official Reference Document to OpenGL, version 1.2 © 1999, Addison-Wesley. OpenGL Architecture Review Board.

6.121   AUDIO LIBRARY SOURCES

OpenAL Programming Guide © 2006, Charles River Media. Eric Lenyel.

6.122   MULTIMEDIA LIBRARY SOURCES

Focus on SDL © 2002, Premier Press. Ernest Pazera.

Programming Linux Games – Building Multimedia Applications with SDL, OpenAL, and Other APIs © 2001. No Starch Press. Loki Games. John R Hall.

6.123   COMPUTER AIDED DESIGN SOURCES

The Art of 3-D: Computer Animation and Imaging © 2000, John Wiley and Sons. Isaac V. Kerlow.

The Blender Book © 2000, No Starch Press. Carsten Wartmann.

6.124   DIGITAL IMAGING SOURCES

Grokking the GIMP © 2000, New Riders. Carey Banks.

6.125   RELATIONAL DATABASE MANAGEMENT SOURCES

MySQL: The Complete Reference © 2003, Osborne McGraw-Hill. Vikram Vaswani.

6.126   ASCII

American Standard Code for Information Interchange. A common character set used in the US and UK computers.

6.127   SVG FORMAT

Scalable Vector Graphics Format. A data format used to store images in a file and to display images on the World Wide Web.

6.128   FBX FORMAT

Film Box Format. A data format used to store 3D models, animations and associated digital data in a file and display 3D models and animations in applications, developed by Autodesk.

6.129   XML FORMAT

Extensible Markup Language Format. A language for describing other languages that describe structured documents stored in a file (e.g. a thesis, an article, a User manual). It was designed to be flexible enough to store and display the huge array of documents on the World Wide Web. But its flexibility means there can be big differences in how it is used between any two documents.

6.130   JSON FORMAT

JavaScript Object Notation Format. A data format for describing hierarchical data structures in a programming language called JavaScript. It was designed to store documents in a file and to display documents on the World Wide Web.

6.131   CSV FORMAT

Comma-separated Format. A data format for describing a Relational Database Table where each row in the table is represented by a line in a file. And the columns in the table are represented by words on each line separated by commas. So for example a 4×3 Database Table would have each row in the table represented by 3 lines. And on each line, the entries in each column would be represented by 4 words separated by 3 commas.

6.132   NEWLINE CHARACTER

An ASCII character which marks the end of a line of text and the beginning of the next.

6.133   ESCAPE CHARACTER

An ASCII character which is reserved for transforming the normal interpretation of the following character in a word. It is normally used to transform a sequence of characters into commands which control how text is displayed. But it can be used to transform a sequence of special ASCII characters (e.g. a Newline character) to ordinary characters in a word.

6.134   X PIXMAP FORMAT

A data format used to hold images displayed on Graphical User Interfaces of computers that use the X Window System.

6.135   CSV FORMAT SOURCES

UNIX (TM) Relational Database Management © 1997, Pearson Education. Rod Manis, Evan Schaffer and Robert Jorgensen.

6.136   SVG FORMAT SOURCES

SVG Essentials © 2002, O'Reilly Media. J. David Eisenberg.

6.137   X PIXMAP FORMAT SOURCES

X Pixmap © 2010, Beta Publishing. Lambert M. Surhone.

6.138   DIGITAL AUDIO TAPE

A magnetic tape used to digitally record music or computer data.

6.139   SECURE DIGITAL CARD

A small portable flash memory card or microchip that stores data in a computer memory, up to 2 gigabytes in size.

6.140   SECURE DIGITAL HIGH DENSITY CARD

A Secure Digital Card that can store up to 64 gigabytes of data in computer memory.

6.141   PULSE CODE MODULATION

A method of encoding an analogue signal in a digital data format. The signal is sampled at a constant rate, and the amplitude at each interval is converted into a number within a limited range.

6.142 DIGITAL RECORDING SOURCES

Desktop Audio Technology © 2003, Focal Press. Francis Rumsey.

6.143 DIGITAL PLAYBACK SOURCES

Modern Recording Techniques © 2001, Focal Press. David Miles Huber and Robert Runstein.

6.144 DATA DESIGN

A description of all the data needed by a game. It is also a description of all the data produced by the tools used to build a game.

6.145 TOOLS DESIGN

A description of all the tools used to build a game. These include the tools used to create the data, to write the computer files used to build the software, to process the data or to archive the data and the computer files.

6.146 GAME WORLD

An imaginary world space in which a game takes place.

6.147 CHECKSUM

A value that represents the total value of a series or sequence of data. That is used to check when there is an error in that series or sequence when it is transferred from one computer or storage media to another.

6.148 LOGIC PATH

Any one of a finite, distinct sequence of actions (or instructions) that can be performed with (or within) a software system (or its software procedures).

6.149 USER MANUAL

An instruction booklet, for software users, which explains how to solve a problem using the software. In computer games, this includes a description of the problem (i.e. the background and goal of the game) and a description of how to use the Interface of the game to solve the problem.

6.150 EXPENDITURE ON DESIGN

An analysis of the relationship between expenditure on design tools and competitiveness of Integrated Circuit companies, conducted by the Electronic Design

Company		1995			2000				2002 (est.)			Ranking	Comments on relative competitive position
		Percent of revenues spent on EDA tools	Importance placed on design capabilities (10 high)	Percent of revenue spent on design tools (multiplier)	Market position within targeted segment	Percent of revenues spent on EDA tools	Importance placed on design capabilities (10 high)	Percent of revenue spent on design tools (multiplier)	Percent of revenues spent on EDA tools	Importance placed on design capabilities (10 high)	Percent of revenue spent on design tools (multiplier)		
IC vendor	Category												
A	Fabless	7.0	9	68.4	1	3.2	9	28.8	3.9	9	35.1	8	Has remained market share leader
B	Fabless	9.4	9	84.6	1	2.7	9	24.3	3.1	9	27.9	8	Has maintained market share leadership
C	Fabless	4.2	9	37.8	3	2.8	9	25.2	3.4	9	30.6	9	Market leader
D	Fabless	4.8	6	28.8	2	2.9	7	20.3	3.3	8	26.4	6	Competitive position strengthening
E	IDM	5.5	8	44	2	3.9	8	31.2	3.1	8	24.8	6	Continues to be strong in segments of the market
F	IDM	2.1	9	18.9	4	1.9	9	17.1	2.6	9	23.4	8	Has become market share leader
G	IDM	1.3	9	11.7	1	1.3	9	11.7	1.9	9	17.1	9	Continues to be market share leader
H	IDM	1.5	8	12	5	1.2	8	9.6	1.8	9	16.2	6	Has strengthened market position
I	IDM	1.1	8	8.8	1	1.3	9	11.7	1.7	9	15.3	7	Has strengthened market position
J	IDM	2.8	5	14	3	2.1	4	8.4	2.5	6	15	6	Lost some market position
K	IDM	1.4	7	9.8	3	1.4	7	9.8	1.5	7	10.5	7	Steady strengthening of market position
L	IDM	1.2	5	6	4	1.3	6	7.8	1.3	5	5	5	Position has steadily strengthened
M	IDM	1.9	3	5.7	2	1.3	4	5.2	1.7	5	8.5	3	Has lost market position
N	IDM	1	4	4	5	1.2	5	5	1.4	6	8.4	6	Positioning strengthening
O	IDM	1	5	5	2	0.8	4	3.2	1.3	6	7.8	5	Has lost market share
P	IDM	1.2	4	4.8	2	1.0	5	5	1.2	6	7.2	4	Has weakened
Q	IDM	1.2	3	3.6	4	0.9	4	3.6	1.3	5	6.5	4	Has lost market position
R	IDM	1.1	3	3.3	6	1.2	4	4.8	1.3	5	6.5	4	Market position has remained weak
S	IDM	1.1	2	2.2	5	1.2	3	3.6	1.4	4	5.6	2	Has lost market position
T	IDM	1.0	3	3	4	0.9	4	3.6	1.1	5	5.5	3	Position has weakened
U	IDM	0.9	4	3.6	6	0.9	4	3.6	1	5	5	2	Position has remained weak
V	IDM	0.9	3	2.7	5	1.1	4	4.4	1.2	4	4.8	2	Position has remained weak
W	IDM	0.8	3	2.4	3	0.9	4	3.6	1	4	4	2	Position has weakened

FIGURE 6.4 A table showing the relationship between expenditure on design tools and the competitiveness of companies that use those tools. (Source: Analysis of the relationship between EDA expenditures and competitive positioning of IC vendors, a custom study for EDA consortium © 2002, International Business Strategies Inc. Jordan Brysk.)

Automation Consortium (EDAC), showed that there was a strong link between the investment in design and a company's future market position (Figure 6.4).

EDA EXPENDITURES AND MARKET POSITIONS OF SELECTED IC VENDORS.

...In general, there is a strong correlation between the importance placed on design implementation capabilities and market position. The IC vendors that placed minimal emphasis on design capabilities in 1995 are not strong in their targeted market areas in 2002, but those that heavily emphasized design capabilities in 1995 have strong market positions in 2002....The analysis of the IC industry indicates that several IC vendors tend to adhere to the same strategies throughout several years, even when it is evident that the strategies have not brought success....

6.151   DRAFTERS

A profession which prepares technical drawings and plans used by production and construction workers. These drawings are used to build everything from manufactured products (e.g. toys) to structures (e.g. an office building).

Drafters held about 213,000 jobs in 2000. More than 40 percent of drafters worked in engineering and architectural services firms that design construction projects or do other engineering work on a contract basis for organizations in other industries. Another 29 percent worked in durable goods manufacturing industries, such as machinery, electrical equipment, and fabricated metals. The remainder were mostly employed in the construction; government; transportation, communications, and utilities; and personnel supply services industries. About 10,000 were self-employed in 2000.

> *Source: Occupational Outlook Handbook 2002-03 Edition © 2002,*
> *The Bureau of Labor Statistics, U.S. Department of Labor*

6.152 COMPUTER AIDED DESIGN (SOFTWARE)

Computer software used to design and simulate physical tests of products which require expensive raw materials before these are physically manufactured, such as buildings, cars and electronic circuits.

6.153 QUALITY CONTROL

A system that accepts or rejects products or services depending on whether these meet all of the customer's specifications and requirements.

6.154 QUALITY ASSURANCE

In theory, a system which ensures that a company's processes (as opposed to their product) will meet all of the customer's requirements and specifications. In practice, software companies just apply two Quality Controls in the latter stages of production, known as Alpha testing and Beta testing, and call this Quality Assurance or QA.

Ensuring that a process would meet a set of requirements implies that the process, as well as its product, would have to be designed. This way you could use the design to test each step of the process, to check that each produces the right result for the next step.

QUALITY ASSURANCE

This is similar to quality control, but has more to do with the process than the product. These are the systems that can demonstrate that the organization can meet the specifications and requirements of the customer. They also allow the management of the organization to know that the customer's requirements are being met.

> Source: The Language of Quality: a glossary of terms and vocabulary
> in plain English. The International Quality Systems Directory © 1999,
> The International Organization for Standardization

6.155 SOAK TESTING

A process for testing a new software or hardware system, to reveal errors that only emerge under extreme conditions. This typically tests the performance of software when many are using it simultaneously and after it has been used for a very long time. This implies that the software has at least three qualities.

Firstly, the process should only be applied to software which could have multiple simultaneous users.

Secondly, the process should only be applied to software whose frequency of use, and duration of use, could be quantified. So that you could compare these figures either to some target or other software which had gone through the same test.

Thirdly, the process should only be applied to software which either has specific targets for its frequency of use and duration of use. Or it has some other existing software whose performance it could be compared to, under the same test. And you use the results to determine whether the performance was good or bad.

But in the Computer Games industry, Soak Testing has been abused on games without these three qualities.

It has been used on games which cannot even have multiple players. It has been used in games where the frequency of use was not and could not be quantified. Since there was no definition of what constituted the use of the game (be it issuing certain commands, completing one stage of the game or completing the whole game). In the context of a Software Evolution Process, with an incomplete game design, all such constituents lack definition.

Finally, Soak Testing has almost always been used on a new game, without any specific targets for the frequency of its use or duration of its use. Nor has the performance of the new game been compared with the performance of existing games, under these tests.

What is Soak Testing?

Soak testing (otherwise known as endurance testing, capacity testing, or longevity testing) involves testing the system to detect performance-related issues such as stability and response time by requesting the designed load on a system.

The system is then evaluated to see whether it could perform well under a significant load for an extended period, thereby measuring its reaction and analyzing its behavior under sustained use. Soak testing is a type of load testing....

Why Should You Perform a Soak Test?

Soak testing is mainly used to identify and optimize potential problems, such as memory leaks, resource leaks, or degradation that could happen over time, to avoid impaired performance or system errors. While stress tests will help the development team to test the system to its limits, soak testing takes the system to its limits over a sustained period of use. In other words, soak testing allows the team to mimic real-world usage, in which the users will constantly need access to the system.

What are the Common Issues that Soak Testing Detects?

- *Memory allocation (memory leaks that finally lead to a memory crisis or rounding failures that only display over time).*
- *Database resource usage (errors when closing the database cursors under certain conditions that would eventually cause the entire system to come to a standstill).*
- *It can also bring about a deterioration in performance, that is, to make sure that the response time after an extended period is as good as it is when the test starts.*
- *Errors when closing connections between tiers of a multilayer system under certain circumstances that could block some or all of its modules.*
- *The gradual deterioration in the response time of some tasks as internal data structures is less organized over an extended test period.*

Source: What is Soak Testing? Learn in 5 minutes © 2024. Katalon Inc.

6.156   WATCH YOUR CHARACTER, IT BECOMES YOUR DESTINY

Quotation from Frank Outlaw. The full quote reads:

> '*Watch your thoughts, they become words.*
>
> *Watch your words, they become actions.*
>
> *Watch your actions, they become habits.*
>
> *Watch your habits, they become character.*
>
> *Watch your character, it becomes your destiny.*'

6.157   QUALITY

The characteristic of a product which meets a customer's needs.

> *The totality of features and characteristics of a product or service that bear on its ability to satisfy stated or implied needs. Not to be mistaken for "degree of excellence" or "fitness for use" which meet only part of the definition.*
> *[ISO8402].*
>
> Source: Quality. The Free On-line Dictionary of Computing ©
> 1993-2001, Denis Howe.

6.158   ROLE OF DESIGN

In industries such as Construction and Electronics, the role of design is not just to describe the plan for making a product, but also to analyse the interaction of the different parts of the plan.

> *Design - the part of the production cycle where creativity, new ideas, ingenuity and inspiration come to the fore. This is also where designers try to model the behaviour of their designs and analyze the complex interactions of millions of constituent parts in their designs to ensure completeness, correctness and manufacturability of the final product. Why? Because it is impossibly difficult, expensive and time consuming to "build it first and fix it later."*
>
> Source: EDA Industry summary © 1998-2002, The Electronic
> Design Automation Consortium

Index

For Product Safety Concerns and Information please contact our EU
representative GPSR@taylorandfrancis.com
Taylor & Francis Verlag GmbH, Kaufingerstraße 24, 80331 München, Germany

www.ingramcontent.com/pod-product-compliance
Lightning Source LLC
Chambersburg PA
CBHW060348220326
41598CB00023B/2846